A Chef in Training – Interview Questions Workbook

A Chef in Training — Interview Questions Workbook

by Chris Flatt

A Chef in Training – Interview Questions Workbook
by Chris Flatt

Orders: Please contact Hospitality Active Ltd, at www.becomingachef.co.uk You can also order via the email address — chrisflatt2003@yahoo.co.uk

ISBN: 978-0-9935196-2-8

First published in 2016

Typeset for Hospitality Active Ltd by Kiryl Lysenka.
Printed in Great Britain for Hospitality Active Ltd by: CMP (uk) Limited, Poole, Dorset.

Contents

Introduction

Hello and welcome to *A Chef in Training – Interview Questions Workbook*. This guide is aimed at commis chefs and chef de parties, working within a kitchen brigade. My name is Chris Flatt, and I have been working as a chef since 1997. Throughout my career as a successful chef, I have learnt what it takes to become successful in the hospitality industry. But what does it take to become a successful professional chef? Some might say that a chef has to be a great cook. This is absolutely true, however there is a lot more to becoming a chef than just being a great cook. Often these other areas are overlooked, and chefs do not fulfil the level of success they deserve. Being prepared for all eventualities will get you ahead of the other candidates who are applying for jobs, promotions and building successful working relationships.

For more information, visit www.becomingachef.co.uk

What Is A Chef?

The majority of people will simply say that a chef is someone who cooks food. While they would be correct in that assumption, the role of a chef is far more varied and important than simply cooking great food. Modern chefs come from a diverse range of backgrounds and experience levels. Despite the greater deal of training that is now required in order to become a modern chef, there are also less obstacles. With the right work ethic, anyone can now become a chef, regardless of age or gender. For example, it is now extremely common to see women working in kitchens, whereas 10 years ago this would be considered unusual. Women are swiftly becoming an integral cog in successful kitchens, and this shows no signs of slowing down.

The kitchen brigade ladder begins with the commis chef, and moves up to a chef de partie. A chef de partie will be running a specific section of the kitchen.

When I first began my chef career, I totally focused on the food I was producing. I put all of my effort into producing great food, learning knife skills, tasting, knowledge of flavours etc. I had many books on all types of foods, sauces and desserts; and loads of recipe books to learn as much as I could. However I never put any thought into other areas that would have helped me to push myself further and faster right from the start. These areas are of huge importance, yet so few people put any real effort into them. This book will cover ALL of these areas, including:

- What it's like to be a chef.
- Chef interview questions and answers.
- The chef interview and working interview explained.
- Communication skills and self-assessment.
- Time management skills and self-assessment.
- Effective team work and self-assessment.
- Organisational skills and self-assessment.

You will also learn skills such as how to prepare for an interview, body language, and how to conduct yourself during a formal interview and working interview, also known as a "stage".

Once again, thank you for your custom and we wish you every success in your pursuit in becoming a professional chef.

Work hard, stay focused and be what you want….

Best Wishes…

Christopher Flatt

CHAPTER 1

Organisational Skills and How to Improve Them

Organisation is essential in a professional kitchen. With this in mind, it's important that every chef is organised, and working towards further organisational improvement. Whether you are a commis chef organising yourself to complete basic jobs (peeling potatoes or putting stock away correctly), or a chef de partie organising his section for a busy service; it's imperative that you get this right.

What Are Organisational Skills?

Organisational skills can be defined as having the ability to use the time, energy and resources that you have available, to achieve the tasks that you have been set to do. Self-discipline is also an effective ingredient in maximising your organisational skills. These skills will help you to plan your work, and implement the procedures needed to achieve your goals.

A chef with good organisational skills is a great asset for any professional kitchen, as they will work systematically to achieve the planned results. Being organised will mean that you are more productive, give a positive impression to the kitchen management, and will be noted for future promotions.

Being Efficient In The Job Role

Being organised will also help you to keep your stress levels down. For example, you might lose time searching for a piece of equipment, or looking for a recipe on a piece of paper. This will leave you in the lurch, and detract from other more important tasks. Organisation = efficiency. It means that you are on the ball, and can complete each task within the allocated timeframe, before moving on to the next one.

Create Good Impressions

When you are organised you will earn respect from your colleagues, and promote your company well by keeping to agreed time frames. Working clean and tidy will also help you to create a positive impression of yourself.

Consider this:

1. The general manager walks past your section and it's a complete mess.

2. The head chef walks past your section and it's well organised, clean and tidy.

How will their opinions differ when they see you working? The 2nd option shows a level of responsibility, and that you are serious about your job and the company. It also shows that you are capable of larger jobs. When a promotion becomes available, you will stand a far better chance of gaining it if you are an efficient, organised and well prepared chef.

Top 8 Organisational Skills:

1. ATTENTION TO DETAIL
This is essential for chefs. You must be able to pay attention to the finer details, and develop the way you work, in order to produce food to higher standards.

2. MULTI-TASKING
Being able to prepare and cook different foods at the same time, especially during a busy service, will garner strong results.

3. ANALYTICAL
When something goes wrong, for example if the oven has broken, you will need to analyse the situation and come up with a logical solution; so that the working day can continue.

4. COMMUNICATION
You must be able to understand people, and able to forward your ideas on to another chef. Chefs are of many nationalities, so communication skills are essential.

5. PROBLEM SOLVING
Having this skill will help you to overcome problems in a systematic way, and not become overwhelmed by difficult situations.

6. DECISION MAKING
You are able to make decisions in an efficient way. This is especially helpful if it's a quick decision that needs to be made during a service.

7. PEOPLE SKILLS
You will need to interact with other chefs, waiting staff or managers in a positive way.

8. TEAMWORK
When working in a kitchen unit, everybody needs to pull together and focus on an end goal.

How To Improve and Develop Organisational Skills

When your organisational skills are developed and you have become more efficient; you will be punctual, reliable and most importantly well-prepared. Your time management will be good. You will take care of your working area and you will also be ready to take on new challenges.

STUDY YOUR CURRENT POSITION
When you're trying to organise a position, determine the systems that are already in place. This could be a new job, or sections you want to improve that you have been working on. Study the current situation, and then develop the organisational aspects of the situation. Firstly, make a list of all the things that will need organising.

START WITH A PLAN

Write down everything that you need to organise and how you will organise this. This could include: putting your recipes onto a computer and into a kitchen file, arranging your equipment so that the most used items are easier to reach, and going over your ideas with a senior manager to check if they are happy/have any input.

USE PRODUCTS ALREADY AVAILABLE

There will be plenty of products already in the workplace to use for organisational purposes. This could include shelves, folders for documents and storage containers. Make sure your files are organised on the computer system so that they can be easily found, for example putting recipes in a recipe folder. Be creative in organising, your section is a work space and not a store for items that are rarely used.

Considerations When Making Your Workspace Assessment

Are there things that you need to change, equipment that is not being used or occasionally used books that would be better in the office? Utilise the space in your storage units, drawers and fridges. Remove any supplies or objects that have been lying around, and be creative with your organisational skills.

Now, complete the following organisational assessment of your work area.

Assessment Page For Your Organisational Improvement

TAKE AN ASSESSMENT OF THE WORK AREA AND HOW TO IMPROVE IT

Areas To Be Improved

1. ....
2. ....
3. ....
4. ....
5. ....
6. ....
7. ....

How To Improve

1. ....
2. ....
3. ....
4. ....
5. ....
6. ....
7. ....

**YOUR WORK SPACE AND HOW YOU WORK
IS PART OF YOUR PROFESSIONAL IMAGE**

CHAPTER 2

Communication Skills and How to Improve Them

What Is Communication?

Communication is the exchange of information between two or more colleagues. Information will be given by one person and received by another.

This will result in a mutual understanding within the workplace, by connecting people who work at different levels and different departments; to achieve set goals.

What Are Communication Skills?

There are three main types of communication skill: verbal communication (speaking), non-verbal communication (non-speaking) and written communication.

SPEAKING (VERBAL COMMUNICATION)

Speaking clearly and using straightforward language will prove essential in getting your point across clearly. Speaking skills are extremely important, and need to be practiced in order to maximise communicational efficiency.

Within the workplace you will need to persuade employees using communication, this skill is crucial when delegating jobs to the staff that you manage.

LISTENING AND READING BODY LANGUAGE (NON-VERBAL COMMUNICATION)

Non-verbal communication is a harder skill to use and needs to be practiced.

You must be able to read a person's body language, and understand the indications that they give off whilst speaking. Developing your non-verbal skills, and having a high level of listening skills, will really help you to gain professional success.

WRITTEN
Written communication is a skill that the whole team needs to be good at, from the lowest level employee to the highest. Your writing needs to be clear, concise and easily readable. There should be no spelling mistakes, and the correct grammar should be used in a professional way. These are important competencies, and will indicate your professionalism.

Why Are Communication Skills Important Within A Professional Kitchen?

We must also remember that within the kitchen, the chefs must communicate effectively with the front of house staff and management. The more effective the communication is with the front of house staff, the better the experience will be for the customer.

Clearly giving instructions to colleagues, and being able to read their body language, will mean the message can be given and well understood. Being able to ask questions, make requests, give instructions and seek clarification are all benefits of good communication. This extends to being able to avoid and resolve conflicts. A good communicator will find it easy to adapt to new situations and read other people's behaviour. This will help you to understand how colleagues interpret your words and actions.

How To Improve Your Communication Skills In The Workplace

BODY LANGUAGE
You will need to display a positive attitude and promote yourself as a reliable and open person. The key ways to do this are: plenty of

smiling, good posture (keep your head up), good eye contact (not too much), open hands and showing that you respect other people's personal space. Good body language will help you to establish trust and rapport. Your colleagues will have more confidence in you, and will be more likely to listen to your ideas.

MAKING PEOPLE WANT TO LISTEN TO YOU

When speaking, it is important that the delivery of your words is clear and easy to understand. Sentences need to be concise and follow a logical structure, making sure that your delivery is flowing.

Speaking openly and confidently will establish trust with your audience; this could be a single person or a group of people. This will help you to gain information from people and make good introductions, along with helping you to build strong relationships with your work colleagues.

BE AWARE OF YOUR SURROUNDINGS

When speaking you will need to adapt your speech to your audience. You should change your choice of words when talking to different colleagues. To be effective at communication, it is important that you are flexible and use language that is appropriate for the listener.

Examples Of Communication Skills

There are many ways in which we communicate at work during the day. This can include conversations with colleagues and suppliers, using the phone and giving instructions. All of our communication needs to be done in the clearest and most effective way possible.

One way of guaranteeing this, is to use the 9 Cs of communication. This is a checklist, which is designed to improve the message we are giving to our audience.

COMMUNICATION WILL NEED TO BE

1. CONCISE

Being concise in your communication will help you to stick to the point and keep it easy to understand. Don't deliver seven sentences when you could have done it in three.

Action points:

- Take out any filler words.
- Remove any sentences that are not needed.

2. CREDIBLE

Highlight your experience and credibility when communicating with a person or an audience that doesn't know much about you.

3. CORRECT

Correct communication will be error free, and will be suitable for the audience.

Action points:

- Do the technical terms fit the chef's experience or knowledge?
- Check all of the spelling and grammar in written work.
- Written communication will need to be readable and easy to follow.

4. CLEAR

When speaking to someone, be clear about your message and what your purpose is in communicating with that person. If you are unclear about what your message is, then the person you are talking to will be unclear and unsure too. This is also the case when writing to someone.

Action points:

- Minimise the number of ideas in a sentence.
- Make sure it is easily readable and the meaning is understandable.

- People should not need to make assumptions on what you are trying to say.

5. COHERENT
This means giving a logical message.

Action points:

- Points are all connected to the main topic.
- Points are all relevant to the main topic.
- The text is flowing and consistent.

6. CONCRETE
Create a clear picture of what you are telling them, for your message to be concrete.

Action points:

- There will be details and facts, but not too many.
- Your message will be highly focused.

7. COMPLETE
The audience has everything that they need to know and understand what they are being asked to do, and all relevant information has been included.

8. CREATIVE
To keep people engaged in your communication, be creative to keep their attention on the message you are delivering.

9. COURTEOUS
When you communicate in a friendly, honest and open way, you are being courteous. You also need to be considerate of the person's views and needs.

**Everyone communicates every day.
The better we communicate, the more credibility we will have
with our colleagues, managers and customers.**

Career Development and Progression

Your career development will benefit from understanding and improving your communication skills. Once you become aware of your weakest areas of communication, it will then be possible to improve upon them. This can be done by taking a personal assessment and putting improvements in place. Communication skills can be learned and improved with practice.

Considerations When Making Your Personal Communication Assessment

By developing and improving your communication skills in the workplace, you will become more successful as a chef. You can then move up to managing others or working as part of a kitchen brigade. When you have the ability to communicate effectively with your colleagues, relations will be improved, resulting in a more effective team.

7 Ways Chefs Can Improve Communication Skills

To help prevent misunderstandings, poor performance and frustration in the workplace, communication needs to be improved. Many of the main causes for work related problems can be traced back to a lack of communication.

1. Be clear and concise.
2. Be aware of body language.
3. Observe others and how to fit in with the team.
4. Don't overreact to on-the- spot questions – think about your response.
5. Learn and use the 9 Cs checklist.
6. Learn to listen and understand what is needed.
7. Be personal, care about people and don't alienate yourself.

Assessment Page For Your Communication Improvement

TAKE AN ASSESSMENT OF THE WORK AREA AND HOW TO IMPROVE IT

Areas To Be Improved

1. ....
2. ....
3. ....
4. ....
5. ....
6. ....
7. ....

How To Improve

1. ....
2. ....
3. ....
4. ....
5. ....
6. ....
7. ....

YOUR WORK SPACE AND HOW YOU WORK IS PART OF YOUR PROFESSIONAL IMAGE

CHAPTER 3

How To Improve Your Time Management Skills

In order for a kitchen brigade to produce meals for customers on time, time management is essential. All of the chefs within the brigade will need to have good time management skills, and should always look for ways to improve upon these skills. If each member of the team is controlling their own time management, the team will work efficiently as a unit, and can produce meals which satisfy the customers' expectations.

What Are Time Management Skills?

Time management refers to the planning and control which is necessary for completing tasks on time. Managing your time well, will result in higher levels of effectiveness and efficiency. When you improve your time management, your standards will go up, resulting in a better standard of work for the team.

Time management will be helped by good organisational and communication skills, which will help you to complete goals on time. For example, completing all of your jobs ready for a busy dinner service.

TIPS TO IMPROVE TIME MANAGEMENT

- Setting the importance of jobs – important jobs and less important jobs.
- Working with the important jobs first.
- Reducing the time spent on unimportant jobs.
- Give yourself a reward when completing set goals on time.

The Importance of Time Management

HELPS YOU SET YOUR TO DO LIST
Once you have created a to-do list, also known as mise en place, you should be able to see which job needs to be done first. Time management will help you to do more jobs in less time. When you have planned your time and know the jobs that need completing in that time, you will work far more efficiently.

HIGHER STANDARD OF WORK
When you know exactly what needs working on, you will become more focused on the job you are doing. The more focused you are, the higher the quality of work that will be produced. A to-do list is also a great way of motivating yourself to do the less enjoyable jobs. These jobs become much easier once you put them down on your list, rather than leaving them open for procrastination.

OUR WORK TIME IS LIMITED
Remember that the working timeframe is limited, so once you have lost some time you will not get it back. This results in lower standards, as you will be rushing and taking shortcuts to complete set jobs on time. In contrast, managing your time efficiently and getting jobs done on time, results in helping yourself to develop self-discipline. This will be extremely useful in your career as a chef.

Key Skills For Time Management

PUT THE MOST IMPORTANT JOBS FIRST, AND THEN WORK DOWN
The goals needed to complete a set task need to be written down in a list, starting with the most important jobs. This will give you a plan, and a set time period in which to achieve all of the goals. A to-do list should be made daily for specific sections of the kitchen, and weekly and monthly for other tasks within the kitchen as a whole.

HOW TO MAKE A DAILY LIST USING THE IMPORTANCE OF THE TASKS:
1. Jobs that are most important and needed immediately.

2. Jobs which are important but not needed immediately.

3. Jobs that are unimportant, and not needed immediately.

The jobs that need to be done will be numbered accordingly, with the number 1 jobs being first on the list, followed by number 2 jobs and number 3 jobs.

WHEN MAKING A TO-DO LIST, YOU NEED TO REMEMBER:

- Don't spend too much time making and managing the list, as doing the set jobs is far more important.

- Don't waste time listing routine jobs on a daily list, have a standard to-do list on your section, maybe in chart form.

- Be flexible in your time management, as you may have a problem and will need extra time to spend on that job.

POINTS TO REMEMBER ABOUT MAKING AND COMPLETING TO-DO LISTS

When making the to-do list, always remember that if a job takes less than 5 minutes, then you should complete it straight away. It doesn't need to go on the list. Always check what needs to be done for the next day and write a to-do list the day before. Whilst you are working on your jobs, turn off any distractions, like a mobile phone. When you have finished your day's work, look back and see where you can improve your time management for the next working day.

Career Development And Progression

Plan your career by using a to-do list. Create a simple to-do list of what you want to achieve in the short term and the long term. Keep your career planner to hand, so you can read it daily. By only making short term goals, there is a danger that you will fall into a pattern of only achieving short term goals, and won't generate any long term achievements. Likewise, short term goals are also important. The immediate future is just as relevant as the distant future. Make sure that you have a 1 year goal, and a 5 year goal. If not, you could find yourself working in a job a lot longer than you had originally planned for.

How To Improve Your Time Management

You will need to spend time planning and organising your working day. If you fail to plan your day, you are in fact planning to fail, and failing to organise yourself in a way that makes sense to you. Set yourself specific goals to be completed on time, and be realistic when doing so. Create a to-do list, listing the jobs that need to be done immediately, and be flexible by allowing time for mistakes that may occur, or interruptions that may happen during the working period.

Try not to be perfect all the time. Attention to detail is good, but paying too much attention to the finer details could delay your over-all career progression. You need to have a realistic approach to achieving goals, in order to achieve your long-term goals within a set time frame. Don't put off important jobs that you have been avoiding, break these jobs up into smaller goals to be completed in set times. Once you focus on the smaller goals; soon the larger job will be completed. Sometimes you just need to say "no". This is hard, but saying yes all the time can have a negative effect on your daily work and overall career progression. Focus on your own goals, and prioritise your friends and family when needed.

When you have achieved something, give yourself a reward – maybe a new piece of equipment – or just a pat on the back.

Improve And Develop Your Time Management Skills

TOP 10 WAYS CHEFS IMPROVE TIME MANAGEMENT

Considerations when making your working day assessment!

1. Being decisive.
2. Noting down reminders.
3. Making to-do lists correctly.
4. Following lists correctly.

5. Being organised.
6. Communicating effectively.
7. Evaluating their performance to make improvements.
8. Understanding cooking times of dish elements.
9. Rewarding themselves for success.
10. Bringing ideas and skills to the team.

Assessment Page For Your
Time Management Improvement

TAKE AN ASSESSMENT OF THE WORK AREA
AND HOW TO IMPROVE IT

Areas To Be Improved

1. . . . .
2. . . . .
3. . . . .
4. . . . .
5. . . . .
6. . . . .
7. . . . .

How To Improve

1. . . . .
2. . . . .
3. . . . .
4. . . . .
5. . . . .
6. . . . .
7. . . . .

YOUR WORK SPACE AND HOW YOU WORK
IS PART OF YOUR PROFESSIONAL IMAGE

CHAPTER 4

Teamwork Skills and How To Improve Them

Teamwork in a professional kitchen is very important, as it takes the whole brigade working together to produce the dishes for the customers. It is crucial that the kitchen brigade has a good leader. The kitchen needs to work together with the restaurant, to give the customers the best experience possible. Everybody has his or her own part to play in the team, in order for the system to run smoothly and effectively. A kitchen brigade can be made up of chefs from many different backgrounds.

Key Teamwork Skills Needed To Become A Chef

To be a part of the kitchen brigade, there are some essential skills that will be needed by all chefs. These skills will make you a valued member of the team and will help you on the path to becoming a successful chef. They include:

ORGANISATIONAL SKILLS

Organisational skills are needed to work effectively as part of a chef brigade. Being organised means that you will be able to get the job done well, and act as a functioning part of the team.

KEEPING CLEAN AND TIDY

All members of the kitchen brigade need to keep their appearance clean and tidy. This is important not only for health and safety reasons, but also so that you are seen as a professional member of the team.

BEING ABLE TO FOLLOW INSTRUCTIONS

Following instructions is crucial to working as a chef, as all of the recipes that are followed are based on a set of precise instructions.

Chefs will also need to be able to take instructions from senior members of staff.

What Are Teamwork Skills?

Teamwork in a kitchen brigade consists of a group of chefs working together to produce meals for the customers. There will be senior chefs on the team who will be leading and running the kitchen, and individual chefs who will be working on their own sections to produce meal components. The kitchen brigade also needs to work with the restaurant staff to fully satisfy their customers' demands.

A good kitchen brigade requires the chefs to work together efficiently. Make sure you understand your role and responsibilities, but understand that these can also be changed from time to time. Actively listen to what other chefs are saying, especially during a service, as communicating at this point is crucial for the team to produce dishes. If you have opinions and ideas, then it is good to express them to senior chefs, but you need to be confident and believe in your opinions.

Be Committed To Your Job Role

Be committed to your assigned jobs and deliver what is expected of you. If you have any issues, ask for some advice from another member of the team. You will be supported by other chefs when you are having issues, but will need to ask for help when necessary. You should also be able to support other members of the team when problems arise.

Commit yourself to learning from the successes you have in your job role, and also learn from your mistakes. When a team is in full flow on a busy service, working together, you will be a valued member of the team. This includes being supported by, and able to support, other staff members.

How Will Teamwork Skills Help Improve The Working Day?

Working together as a team will accomplish a lot more than an individual chef can by working on their own. Different chefs within the team will have their own skills, opinions and experience to bring to the brigade.

The Important Teamwork Skills Needed

Good communication is essential in developing a productive work environment. If you do not communicate correctly, this can lead a breakdown in the team, which will reflect badly on everyone in the kitchen. When you are working on a section, make sure that you understand the tasks you are required to do. This requires both you and your manager to communicate effectively. If you are not clear, ask for a more thorough explanation of your job duties.

Once you have been working in a kitchen brigade for some time, and have a good knowledge of how things work, it is important that you do not develop an ego. Concentrate on your own development as a professional chef, rather than telling other people how things should be done. If you have opinions, you can express them to your managers. Once you are in a senior position, you will be able to use your knowledge to develop other chefs within YOUR team.

You will also need to develop relationships with other chefs, in order to help the team grow. By eliminating misunderstandings, which could lead to conflict, you can ensure that the kitchen brigade is working in unison. Remember that everybody in the team will have unique qualities and a different background. As professional chefs, we need to respect other views, religions, genders and opinions. The diverse range of people working in a kitchen actually helps when finding solutions to problems. Each chef will have their own skills, knowledge and opinions. This means that you'll have a whole variety of solutions to consider.

Finally, remember that it's good to be competitive, but not too much!

Competing with other chefs who have succeeded on their sections, or have created a great dish that will go on the menu, is good for your motivation. Try and do even better than them! This is a great way of motivating yourself to improve your own skills, and is also good for the team, as it means that the overall standard of work will improve.

Career Development And Progression

When planning for your future success, you will need to take the time to learn leadership skills. This means taking control of a section, and then a kitchen. First you will lead individuals, and then a team of chefs. To be a successful leader, you will need to develop certain personal skills. These will include:

CONSCIENTIOUSNESS
Being efficient and organised.

AGREEABLENESS
Being kind, sympathetic and cooperative.

INTELLIGENCE
Using the knowledge you have learnt productively.

ENERGY LEVELS
Being motivated with energetic ora.

STRESS TOLERANCE
Adapting well to stressful situations.

SELF-CONFIDENCE
Believing in yourself and your abilities.

EMOTIONAL MATURITY
Having the ability to understand and manage your emotions.

Top 10 Skills Needed By Chefs For Teamwork

1. Following instructions.
2. Being organised.
3. Good communication.
4. Good time management.
5. Working clean and tidy.
6. Respecting others.
7. Develop relationships.
8. Committed to the team.
9. Learn from failures.
10. Be motivated.

Assessment Page For Your Teamwork Improvement

TAKE AN ASSESSMENT OF THE WORK AREA AND HOW TO IMPROVE IT

Areas To Be Improved

1. . . . .
2. . . . .
3. . . . .
4. . . . .
5. . . . .
6. . . . .
7. . . . .

How To Improve

1. . . . .
2. . . . .
3. . . . .
4. . . . .
5. . . . .
6. . . . .
7. . . . .

YOUR WORK SPACE AND HOW YOU WORK IS PART OF YOUR PROFESSIONAL IMAGE

CHAPTER 5

Memory Skills and How to Improve Them

A good memory is an essential part of being a chef, and a key area that can be improved upon to get yourself ahead in a professional kitchen. A great memory is needed for remembering recipes, dishes, timings of different foods, correct cooking temperatures of different foods, correct cooking methods of ingredients to produce great quality dishes and getting orders correct during a busy service. This helps to prevent wastage and keeps standards high. You will not always have time to refer back to books or checking online recipes, and therefore improving your memory is a vital aspect of training as a chef.

Working in a professional kitchen can be tiring and stressful, which itself can cause memory loss. Knowing how to improve your memory will be a massive advantage.

What Are Memory Skills?
Memory is how we remember information when needed.

There are three stages of memory to improve upon:
- How we take in the information and understand it.
- Storing the information in our memory.
- Recalling the information when needed.

Having memory loss in a busy professional kitchen may cause problems. Memory skills are used for remembering orders, recipes, prep lists and remembering the timings of foods. When taking instructions and running a section, you will need to remember the goals set and the orders being called. During a busy service this can be a challenge, however memory can be improved through practice.

Top Memory Improvement Related skills

To improve your memory, there are some skills that can be developed by practicing. These skills include:

FOCUS

Focusing on the task being undertaken, rather than multi-tasking too much, is a great way to improve your memory. If there is something important that you need to remember, stop multi-tasking and concentrate on this. This will help you to store the information in your memory for when it's needed.

TASTE, SMELL, TOUCH AND SEE

The more senses used when taking in information, the better. For example, when learning to cook a new dish. By tasting, smelling, touching and seeing the dish; the kitchen chefs will develop a good understanding and memory of the dish when preparing the ingredients. The more senses used when remembering a piece of information, the stronger the memory will become.

CHUNKING INFORMATION

This method of memorisation involves breaking down a large amount of information into smaller chunks, and then focusing on remembering the individual chunks. This can be used when memorising a sections prep list. If there is a long prep list to go through, put the items into chunks. This could include: garnish, meats, and sauces.

ORGANISING INFORMATION

Chunking helps to organise our brains, and our brains like being organised. The more organised the information, the better. Our brains will take in this information, store and recall it.

USING SUGGESTIONS AS CLUES

This method involves using imagery, rhymes or songs to remember information. For instance, in a professional kitchen there are 6 colour coded chopping boards. We can associate these colours with images, to make things easier to remember. For example, a blue board is for raw fish. I relate the colour blue with the colour of the sea, which brings me back to fish.

LEARN HOW IS BEST FOR YOU

At the end of the day, there are many different ways to remember information. Whether it's writing, rhyming or taking photos; you just need to learn the best way for you.

How To Improve And Develop Memory Skills

Improve Your Lifestyle – Improve Your Memory

EAT HEALTHY

Having a good diet is great for your memory. There are certain foods that can be included into your diet, which specifically accelerate the process of remembering information. Foods specific to memory improvement include celery, broccoli, cauliflower and walnuts. These foods all contain compounds which protect the health of your brain. It's essential that you get plenty of fresh vegetables and healthy fats into your diet.

EXERCISE

Exercise is also very important. It helps you to keep yourself healthy, and is also a way to help keep the brain healthy, by strengthening cells. When you exercise, chemicals are released that promote the brain's health, leading to improved memory.

AVOID MULTITASKING

Multitasking may make you forgetful, slow you down and possibly lead to mistakes. Getting a piece of information into the memory takes about 8 seconds. Let's say you are talking to a staff member whilst carrying a box of ingredients and put down your locker key. You may not remember where you have left your key. If you had completed the process of moving the ingredients to your locker, without any distractions, you would be far more likely to remember where you had placed it.

INTERESTS

Learning a new skill that you have an interest in, and is challenging, will help exercise the brain. This will significantly help with improving your memory. Find an interest that is stimulating and holds your attention, whether this is learning to play an instrument or growing your own foods; there are almost unlimited choices.

Tests To Train And Improve Memory

BRAIN GAMES
The brain needs to be challenged with new information, or it begins to slow down and becomes less healthy. A great way to challenge your brain is by playing brain games. There are websites available that have games specially designed for brain exercise.

Brain Games Website - www.brainhq.com

3 More Tips For Brain Health

VITAMIN D
Vitamin D is an important vitamin for the brain. It helps with planning, processing of information and making new memories. Relevant sun exposure is all that is needed to keep your Vitamin D levels healthy.

INTERMITTENT FASTING
The best fuel for the brain is ketones. This can be generated by not eating carbs, and introducing other healthy fats into the diet. A one day a week diet can help the body to reset itself, and will burn fat instead of sugar. This diet also decreases calorie intake, which is also good for your brain's health.

KEEPING YOUR GUT HEALTHY
Your gut is also known as your second brain, and your gut health can have a huge part to play in your brain function and behaviour. The gut transmits information to the brain through a main nerve. Avoiding sugar helps to support gut health, as does taking probiotic supplements. The latter contains healthy bacteria, which are great for the gut.

Considerations When Making Your Assessment

Can you remember all of the cooking methods, ingredients and techniques needed for a menu? Are you struggling with the cooking time of foods? Are you good at remembering all the customer orders on a busy service? Do you miss items from your list when ordering goods?

Assessment Page For Your
Memory Skills Improvement

TAKE AN ASSESSMENT OF YOUR MEMORY SKILLS
AND HOW TO IMPROVE

Areas To Be Improved

1. . . . .
2. . . . .
3. . . . .
4. . . . .
5. . . . .
6. . . . .
7. . . . .

How To Improve

1. . . . .
2. . . . .
3. . . . .
4. . . . .
5. . . . .
6. . . . .
7. . . . .

YOUR WORK SPACE AND HOW YOU WORK
IS PART OF YOUR PROFESSIONAL IMAGE

CHAPTER 6

How to Complete an Application Form and CV Essentials

Applying for a job doesn't always mean sending off a CV and a cover letter. When starting the recruitment process, some employers ask only for an application form. In this section of the guide I will be giving advice on how to complete the application form.

The application form is the first point of contact that an establishment will have from you. Therefore, it's important that your application form is concise, accurate and makes a good impression. You also need to make sure that all of the relevant sections have been completed. If a question is not relevant to you, then write N/A in the answer. Some employers will have application forms that need completing online.

When you are applying for a job at any level, you will be competing against many other applicants. With this in mind, it is vitally important that the form is completed to the best of your ability. Think of it like this:

- An employer has had a very busy week, spending hours reading application forms and now he comes across your application. Your application form is hard to read, full of grammar errors and is incomplete in a few sections. Do you think he will be in contact with you for a follow up interview?

Many jobs will require you to fill out an application form, **as well** as a cover letter and a copy of your CV. However if you have only been asked to fill out an application form, don't send a CV, as this will double up the information.

Employers will use application forms to find specific information about you that may not be detailed on the CV. Application forms are generally used when employers are expecting a lot of applicants. For this reason, the application form needs to be completed to a high level.

How to make your application form stand out

Always read and go through the application form thoroughly, before answering any questions. Here are some tips on how to make YOUR form stand out:

- Print a spare copy of the form so that you can practice, and limit the number of errors. Application forms can be photo-copied, or if online you can download extra copies.

- Make sure that you use the correct coloured ink, and that your handwriting is easy to read and very neat.

- Keep your achievements relevant to the requirements listed in the job description.

- Have your form read through by a trusted person, and ask them for an honest opinion. Always read through and check for grammar and spelling errors before sending the form off. Make sure you keep a copy of the completed application form, so you can make reference to it during an interview.

- To prevent any worrying about the form (if it needs to be sent by post) it can be sent by recorded next day delivery for a little extra money. This will put your mind to rest, and you will have confidence that the form will get to the employer promptly.

If the application is being sent off by post, it is a good idea to include a cover letter, so that you can give a brief introduction of yourself.

APPLICATION FORM TIPS

Completing an application form should be easy, but many people are in the habit of rushing and will end up making mistakes. Taking the time to complete the form correctly is essential, as this will be the first part of the selection process. If you have spelling mistakes, use the wrong ink colour or miss a question, then you are setting yourself up for failure.

MORE ADVICE AND TIPS

You will need to be able to provide evidence of where you meet the job requirements. Give relevant details of your responsibilities, strengths and skills wherever it is possible, but make sure you can back these up. Go through the job specification and identify the requirements for the job, before trying to implement these key requirements in your form.

A good method is to try and identify the keywords and phrases used in the job and person specifications. Use these as a guide for your answers to the application form questions.

Always remember to follow the guidance notes that come with the application form. If the form asks you to use black ink then do so. Your application could be rejected for using the wrong coloured ink. Fill out your personal details honestly. If the employer notices any false details, then it may come back on you at a later date and result in disciplinary procedures. This can also make your contract void.

A Blueprint For A Successful Application Form

ACTION POINT 1

Get 2 copies of the job application form, a copy of the job description and the person specification. When you have these crucial forms, find a highlighter pen.

ACTION POINT 2

Now it is important to read through the job description and identify the key requirements, highlighting these areas with your pen. Examples of these may include:

As a chef de partie, you will need to be able to:

- Deal with suppliers, orders & deliveries.
- Set up, cook, and clean for the evening meal, to the highest quality.

The successful chef de partie candidate will:

- Be self-motivated with a very positive attitude.
- Have Health & Safety, Food & Hygiene (NVQ2) certificates.

ACTION POINT 3

When you have identified the key requirements of the role, your next job is to give evidence of when you have met these requirements. For instance, if one of the key requirements is 'to deal with suppliers, orders & deliveries', you should give an exact example of where you have done this in a previous role.

ACTION POINT 4

It is good practice to follow up your application form with a phone call, a couple of days after it has been sent. Try not to call too soon or repeatedly, as you may come across as enthusiastic but too pushy.

CV Essential Tips

A CV is a summary of your career history. This is an extremely important document, as it will demonstrate the skills and experience you have gained, and why these are suitable for the job role that you are applying for. An effective CV will match the specifications of the personal description within the job advert.

Do not send the same CV to every job that you apply for. Every CV you write must be custom built around the description of the job advertised by the employer. When you are reading a job advert, look out for the keywords being used to describe the skills and experience needed. For instance, if "customer focused" and "excellent craft skills" are used in the job advert, make sure you clearly state in your CV that you are "customer focused" and have "excellent craft skills".

How to structure your CV

Start your CV with your personal contact details. This includes your name, phone number and your email address. When using your email address, make sure you are using a professional address. For instance, a contact email that reads xsaucyemmaxx@ domain. co.uk is unprofessional and should not be used. A good tip would be to use your name, for instance: chrisflatt@domain.co.uk.

Your personal profile statement

Following this, you should insert a personal statement that is tailored to the job role in question. Once you have identified the keywords you need to use within the statement, you should try to answer the following questions.

- Who are you?
- What do you have to offer?
- What are you aiming for in your career?

Employment history

When writing this section, start with your latest job and then list your previous jobs in reverse chronological order, along with the dates and the position you held. Give a truthful run-down of any specific responsibilities that you had. I recommend that you write down any skills you used, that the employer is looking for in the job description.

If you have no experience at all, you should aim to gain as much unpaid work experience as possible before applying, so that you can list this on your CV.

Education

Start with the dates studied and the school / college you attended. List your most recent achievements first, and then list your qualifications in reverse chronological order. You will need to include education from GCSE or equivalent, and any other qualifications that you have which are relevant to the job that you are applying for.

Hobbies and interests

This is optional, but if you lack any real work experience then your hobbies are a good opportunity to put your personality across. Be careful not to use anything too generic such as "socialising with friends". Try to make your hobbies relevant to the job you are applying for. For example, "training at the gym" is good for a chef, as it is a physical job. Be prepared to expand upon your interests during the interview if called upon.

References

It is important to have 2 good references included as part of your CV. Make sure that you have asked your referees for their permission beforehand, so that they are aware you have used them. It is possible to make references available on request, but it looks a lot more professional to have the reference details stated on the form.

How is a great CV created?

First of all, let's take a look at an example job advertisement. Let's assume you are applying for a chef de partie job, which is advertised as follows:

Example job advertisement for a chef de partie at a 4 star hotel in London:

We currently have a new opportunity for an experienced chef de partie to join our team. This is an excellent opportunity for a customer-focused individual to join a leading food and management company, which offers great opportunities for career progression.

You will be responsible for:

- Preparation and cooking of food;
- Ensuring the kitchen areas are clean and tidy;
- Ensuring health and safety regulations are followed at all times;
- Delivering a first class customer experience;
- Assisting in all areas of the kitchen for breakfast, lunch and dinner;
- Assisting the Head Chef or Chef Manager with paperwork, including stock ordering, menu planning and menu costing.

The successful candidate for this role will have excellent craft skills with previous chef experience and good communication skills. It will be advantageous but not essential to have a basic Food Hygiene Certificate and a basic Health and Safety Certificate. The successful candidate will also have good customer service skills.

This is a great opportunity to work in a fast paced kitchen with a very experienced head chef, and there will be ample opportunity to progress your career.

Please send your CV to: chefdepartiejob@example.co.uk

Key requirement keywords identified:

1. Experienced chef de partie;

2. Customer-focused individual;

3. Good communication skills;

4. Food Hygiene Certificate and a Basic Health and Safety Certificate;

5. Fast paced kitchen;

6. Good customer service skills.

How to tailor your CV to the job description

First of all, you need to identify the key requirements listed above, and build your CV around these. If applying for various positions within different companies, this can be extremely time consuming. A tip for this is to focus on applying only for a set number of positions at any one time. Set yourself a limit, and don't go over it. Applying for too many positions could lead to mistakes. Your CV has a much higher chance of going through the initial stages if it is concise, matches the key requirements and is easy to read.

To follow are 2 sample CV layouts for a Chef de partie:

SAMPLE CV – 1

Curriculum Vitae

Name here
Address here
Email: namehere2003@domain.co.uk
Tel 07899 XX 87XX

PERSONAL STATEMENT
I am a hard-working, customer focused chef de partie, who is always looking to develop my excellent craft skills. I thrive when working under pressure within a fast paced environment.

I am a very clean and tidy chef, and always follow health and safety guidelines. I enjoy working in all areas of the kitchen, assisting with breakfasts, lunches and other hospitality requirements. At all times, I am happy to assist the head chef in any other areas which make for the smooth operation of kitchen tasks. Finally, I'm in possession of:

- An up-to-date basic Health and Safety Certificate;
- An up-to-date basic Food Hygiene Certificate.

EDUCATION AND QUALIFICATIONS
2008 – 2009 – Cambridge Regional College - B-TEC pass in electrical engineering.

2003 – 2008 - Soham village college GCSE passes in maths, religious education, craft design and realisation.

PROFESSIONAL EXPERIENCE
2013 – Present: Chef de partie at 4 star hotel, fast paced 1 rosette restaurant. Working on all sections of the kitchen including some breakfast shifts when needed.

2012 – 2013: Demi-chef de partie at 4star hotel, 1 rosette restaurant and busy banqueting business. (Excellent craft skills learnt in this position).

2009 – 2012: Brasserie style restaurant, commis chef. Completed NVQ level 2 in Professional Cookery whilst working here.

INTERESTS AND HOBBIES
Travel, skiing, running and reading cookery books.

REFERENCES
1. Mr xxxx Head Chef at sample hotel.
2. Mr cccc Head Chef at other hotel

....

This example of a CV (Sample CV 1) has been written using the previous job description as a blueprint. It is well written, tidy and compact; and provides skills that are specific to the job. I strongly recommend that you pay close attention to the way I have used keywords in my CV, in order to match the job description.

....

SAMPLE CV – 2

Curriculum Vitae

Your Name
Your Street Address
City, county, postcode
Mobile number 07899 xx 87xx
Email: namehere@domain.co.uk

PERSONAL STATEMENT

I am a hard-working, customer focused chef de partie with excellent craft skills, which I am always looking to further develop. I thrive when working under pressure within a fast paced environment. I am a very clean and tidy chef, who always follows health and safety guidelines. I enjoy working in all areas of the kitchen, assisting with breakfasts, lunches and other hospitality requirements.

EDUCATION

First school attended:

Soham Village College
GCSE Passes in maths, religious education, craft design and realisation.
Soham
Graduated July 2008

Any other school:

Cambridge Regional College
B-TEC pass in electrical engineering.
Cambridge, Cambs
Graduated July 2009

EXPERIENCE

4 Star hotel   London   Apr 2013 – Present

Position: Chef de partie

- Working on all sections of kitchen;
- Breakfast shifts when needed;
- Fast paced restaurant.

4 Star hotel   London   Apr 2012 – Apr 2013

Position: Demi chef de partie

- 1 AA Rosette Restaurant
- Excellent craft skills learnt here;
- Working on larder and sauce;
- Worked in busy banqueting when needed.

Brasserie style restaurant   London   Apr 2009 – Apr 2012

Position: Commis Chef

- Competing NVQ level 2 in professional cookery whilst working here.

INTERESTS AND HOBBIES
Travel, skiing, running and reading cookery books.

REFERENCES
1. Mr xxxx Head Chef at sample hotel
2. Mr cccc Head Chef at other hotel

....

Downloadable CV Templates @: www.becomingachef.co.uk

Tips to help you write a winning CV:

Read the job description carefully, so that you have a clear under standing of what is being asked for. When you read it, highlight the keywords you have identified.

Do not pad your CV out, keep it easy to read, precise and relevant to the job that you are applying for. A lot of white space is good for a CV, as it makes it easier to read. Remember quality is more important than quantity.

Chris's CV tips:

- A CV needs to be no more than 2 sides in length;
- Check spelling, grammar and punctuation. You can ask for a proofreading of your CV to gain a second opinion;
- Use keywords from the job description and include them in your CV;
- You need to create a positive profile of yourself.

Writing a cover letter

A cover letter needs to be attached to every CV that you send off to employers. This is your first chance to make a good impression, and a lot of thought and preparation needs to go into the letter. The cover letter needs to be relevant to each job that you are applying for, and needs to relay how your skills are relevant to the job.

Chris's cover letter tips:

Make sure that there are no grammar, spelling or punctuation mistakes. Ask a trusted person to proofread the letter several times, do not use abbreviations. Remember to use keywords and key phrases you have taken from the job specification and description.

Your CV and cover letter will also need to be strong. For more information, go to www.becomingachef.co.uk and use our tips to help you succeed.

If you follow our blueprint for a successful application form, CV advice and meet the key requirements for the role (detailing your specific evidence), your chances of successfully passing this stage will greatly increase.

CHAPTER 7
The Chef Interview

The chef interview is usually a two-step process, and both of these steps are extremely important. The first interview in the process will probably be the most nerve wracking. It is perfectly normal to feel nervous at this point. I believe that this will actually help you to perform to your best.

Remember that once you have reached this point, your application and CV have both already been accepted. This means that you are in good stead. The interviewer has asked you for an interview, as they believe you could be a good fit for the job, and they want to find out more about you. You have their attention.

HOW LONG WILL THE INTERVIEWS LAST FOR?
The first interview will usually last between 45-50 minutes. The second interview, also known as the working interview, will last between 2-3 hours.

HOW SHOULD I DRESS FOR THE INTERVIEW?
It is not essential that you wear a suit, as the interviewer will not assess you on your clothing. However, it is highly **recommended** that you wear a pressed and clean suit, with clean shoes and no silly socks or ties!!

WHY DO WE HAVE AN INTERVIEW?
Primarily the interviewer will talk to you about your interest in the job role. The interviewer will then delve into some of your experience, to see if you are a good fit for the job.

WHEN WILL MY WORKING INTERVIEW TAKE PLACE?
The working interview can take several different forms. It's imperative that you find out what will be taking place before you attend the interview. Sometimes the interview will take place straight after

your first interview, but it may also take place on another day. Check before you attend the first interview, and don't forget your uniform and equipment if necessary.

How To Prepare For The Interview

When you're applying for jobs, it's fairly easy to find out about the company you are applying to, and the people who own and run it. If you are applying to a business directly, you can go to their website and pick out as much information as you can. This will help you in your interview preparation. If you are applying through an agency for a chef's position, you will need to ask the agency for as much information about the company as you can. The agency will not give you the name of the company at this point. You should also think of some questions to ask the interviewer about the company, as this will show your interest in the job role. Whilst you are going over the website, make sure you look at the menus and get a good feel for the style of food. Go through their galleries for more ideas about the venue and the food it serves.

Research

You need to make sure you know the area and how to get to the interview. It is advisable to plan your journey well. If you are travelling by public or private transport, it is always a good idea to test the route before the interview day. You should set off ahead of time, and plan on being at the interview ten minutes early. Organise everything that you need to take with you, the day before you attend. This may include a copy of your application form, but always make sure you take a copy of your CV and your references too.

How To Dress

At the chef interview it is very important to make a good first impression. You need to look clean, neat and smart. It is best to always wear a suit to the interview. Make sure it has been cleaned and ironed and wear a pair of polished shoes. You will not lose a job for being too smart at an interview, but you may lose marks and make a bad impression if you go to an interview casually dressed. Being casually dressed can come across as lacking a serious attitude, and will reflect poorly on the employer.

More Tips For The Interview

Your chef job search has now been completed, and you are ready to meet with the employer and attend an interview. Here are some extra things to remember. In order to shine during your interview, you will need to be confident and have a strong belief in yourself. It is perfectly normal to be nervous before an interview, but do not let your nerves get the better of you. Keep them under control. The more confident you are, the less the nerves will impact you. Confidence comes with being prepared and well-practiced. The employer has asked you to come in because they believe you are a good fit for the job, and are capable of doing the job to a high standard.

Your job now is to back up the first impression given by your outstanding CV. You will need to practice your interview technique, so that you do not get stuck for words. Practicing what you are going to say and how you will present yourself is very important. In order to make sure that the interview runs as smoothly as possible, the areas you will need to pay attention to include:

- Do your homework.
- Create a great first impression.
- Be aware of good body language.
- Be confident in all answers.
- Talk positively.
- Remember the interview is a two-way process.
- Finish confidently.

The Interview

As mentioned, a chef's interview usually consists of a two-part process, although you will need to pass the first interview in order to secure a working interview. I will focus on the second part later in this chapter..

How To Conduct Yourself In An Interview

During the interview you need to conduct yourself well. This is known as having good interview etiquette. In order to master this, you will need to practice your interview techniques beforehand.

BE PREPARED TO HAVE A CONVERSATION

Once you are settled in the interview and the preliminary questions have been answered, you will have opportunities to talk about how your work can benefit the company. Providing you have researched and practiced, you should be ready for this. Have your interview answer ammunition ready to give the interviewer, but keep some back for later, like a secret weapon that will really impress them.

Remember that the business could be in a great position, so you need to show how you can help the business with its continued success. If the business has some bad points, show how your specific skills will help them. Look out for clues that the interviewer might give for areas in which their business needs to improve, and show how you can add positivity to this.

Interview Etiquette

- When the interviewer comes into the waiting room, make sure you are sitting upright but not rigid and not slouching. Stand up and give them a good handshake, practice your handshake beforehand to get it right. Make sure you are not chewing any gum and turn off your mobile phone.

- When you walk into the interview room, don't sit down until you are invited to do so. Introduce yourself confidently and politely.

- If you are offered a glass of water or a hot drink, it is good practice to say yes.

- Good eye contact is important but you do not want to stare at the interviewer too intensely. Be polite by saying please and thank you in your interview.

- Always listen to a question until the interviewer has fully completed it, and do not interrupt. This comes across badly, and there is a risk of giving the wrong answer. Listen to all the questions fully, and then answer them directly and precisely.

- It is possible that the interviewer will take you for a walk around the establishment. Always walk beside, and not in front of them, as well as holding the door open for them. Be aware that there may be guests around. If a guest is behind you, it's good practice to also hold the door open for them too.

- During the interview there may come a point where the interview is less structured and feels more relaxed. At this point be careful, you do not want to start talking about your previous employers and colleagues badly, and giving off negative energy. You should start by talking about any skills and professional attributes which will be of value to the interviewer. Having done your research, and practiced your interview technique, you should be confident in this.

- A good sign is when the interviewer starts to talk about themselves and the company. At this point, make sure you listen and encourage them to keep talking.

Thinking About Salary

During the first interview, it is good practice not to bring up your salary expectations. There is a possibility, however, that the employer will; and you need to be ready for this. Know what salary you are looking for, but make sure your lowest range is not higher than their highest range. You should have a feel for how well the interview is going by now, and as long as you have established your value to the interviewer, you could go in just below their highest. This will show that while you value yourself, you are willing to compromise and meet their expectations.

Before You Leave

As the interview comes to a close, make sure you have an understanding of the next stage. Will they contact you or do you need to contact them?

Once the interview has finished, make sure you shake hands with the interviewer. Be aware of your speech, do not use slang words and never swear.

Body Language Dos And Don'ts In The Interview

HANDSHAKE

The handshake needs to be dry and firm. A limp sweaty hand is very bad. This is the first point where your body language is being watched.

EYE CONTACT

Have good eye contact but do not stare at the interviewer. If you find it difficult to look the interviewer in the eyes, look at their nose instead.

HANDS

It's ok to use some hand gestures, but don't overdo it whilst you are talking. Practice responding to interview questions in front of a mirror, so you understand how you look when responding.

POSTURE

Your posture shows your energy, self-control and your enthusiasm. Stand and sit upright at all times and never slouch. Slouching reflects a negative attitude in an interview. Don't cross your arms, lean forward anxiously or click your fingers. You should also try to avoid tapping your feet.

RELAX

Be relaxed, avoid touching your hair, tapping your fingers or playing with pens. The interviewer will be watching your posture, your gestures and the facial expressions you use. It is good to smile at appropriate times.

The Working Interview

Once you have had a successful initial interview, you will be asked to attend the second part of the interview process. In some cases, you will be expected to attend the second stage of the interview directly after your sit down interview. Therefore, it's important to make sure you know what this interview will involve. If the practical test is straight after your sit down interview, then you need to be prepared by bringing your chef whites and knives.

THE FOUR MAIN TYPES OF WORKING INTERVIEW

A SECTION WORKING INTERVIEW
This type of interview involves the applicant working a shift, on the section he has applied for. The applicant will be working with the current chef running the section.

A TRADES TEST
This test often involves the applicant cooking a classic dish. This is usually a French classic that the applicant needs to demonstrate his skill with. From a trade test, a chef can see how you work and watch your skills.

THE MENU TEST
This test often involves the chef cooking a three-course meal. He will create the menu using the produce within the kitchen. Remember to cook dishes in line with the style of cuisine at the establishment.

INVENTION TEST
This test involves the chef cooking a mystery box of ingredients. The chef will be looking for your creative skills and how you use the produce.

I will go into this in more detail further in this guide

The S.T.A.R. Method

The S.T.A.R. method is an effective technique for answering difficult interview questions. This method will help you to structure a positive answer to situational interview questions. Many of the questions will require the candidate to not only answer the interview question, but to give evidence of where they have implemented the skill that is relevant to the chef's role.

Below are two examples of the type of question that you may be asked, where the STAR method can be applied:

"Can you give me an example of when you have worked as part of a team?"

"Have you ever been asked to train a member of the team?"

It is very easy to say —

"Yes I work as part of the team every day, as a chef de partie on the starter section."

Or

"Yes, I am often asked to train new chefs on a section when needed without hesitation."

Anybody can give answers like this, but we want to stand out and give a structured answer, with evidence to prove what we are saying. This is where we will use the S.T.A.R. method.

The S.T.A.R. Method

S - Situation
T - Task
A - Action
R – Result

1. SITUATION

To respond to the interview question, we will start with the situation. First of all explain the situation, and how you and other people were involved.

2. TASK

Now explain the task that you and the other people were required to do.

3. ACTION

Now explain the action that you and the other people took. It is crucial that you explain your actions, be specific about the action and make sure your actions stand out from those of others. Within the answer, the key evidence needs to be detailed, and you need to explain how your actions directly helped the overall result.

4. RESULT

It is recommended that the result shows your actions in a positive way. So the action you took directly helped to complete the task that resulted in a positive outcome.

A tip at this point is to add a sentence
on how you could improve upon the positive result,
for the next time the situation arises.

An Example Answer To An Interview Question Using The S.T.A.R. Method

Q. "Can you give me an example of when you have worked as part of a team?"

Situation – The situation I found myself in was that I was about to finish my morning shift, at 3pm. The chef who was to take over my section was going to be a few hours late, due to his car breaking down. I was asked to cover until the chef arrived, as this was a busy day.

Task – It was mine and the team's responsibility to prepare the section, get ready for a busy service and finish the day when the chef arrived.

Action – The action others took was to prepare the section whilst I had a break. I then came back early and finished the preparation. I also completed a work list and made the orders for the following day.

Result – The chef was unable to make it to work, so I stayed for the evening service and was responsible for the section. This resulted in the kitchen running smoothly and the team working well on a busy evening service.

If this situation happened again, I would suggest the chef takes the day off, especially if they will be more than an hour late on a busy day.

It is recommended to practice using the S.T.A.R. method. Write down a list of questions and your answers, and then practice reading them. Employers are looking for good communication skills. If you practice thoroughly, you will be able to demonstrate this quality to the interviewer.

During the interview stages, the interviewer will be assessing you on the following areas:

BACKGROUND
Asking questions about your previous jobs and hobbies, checking for any red flags.

YOUR INITIATIVE SKILLS
The interviewer is looking for examples of where you have done specific tasks, without being told to do so.

YOUR STRESS TOLERANCE LEVELS
Checking that you have previous experience of performing under pressure.

YOUR PLANNING AND ORGANISING SKILLS
Looking for an ability to plan work and handle multiple tasks.

SPECIFIC TECHNICAL QUESTIONS ABOUT THE JOB POSITION
Checking for previous responsibilities related to the job being applied for.

YOUR WORK STANDARDS
Checking for high personal standards of performance.

YOUR TEAMWORK SKILLS
Looking for a strong ability to get along and work with others.

YOUR COMMUNICATION SKILLS
The interviewer will be looking for you to be able to express yourself in an effective organised way.

LEADERSHIP SKILLS
Looking for your skills and methods of guiding other people.

JOB MOTIVATION
The interviewer is checking for areas that you enjoyed the most in your current role, and if those specific areas are part of the available position.

CHAPTER 8

Sample Interview Questions and Answers

The two main types of questions:

- Motivational questions
- Situational questions = STAR Method

During the chef interview you will normally be asked 10-15 questions. Be prepared and practice your answers, a good tip is to set up a role-play with a friend or family member before taking the interview. Remember that the questions are not guaranteed to be the ones you get asked at the actual interview.

1. Tell me about yourself?

This is a great warm up question, and is usually the opening question in most interviews. In your response to this question, you need to use keywords. The keywords to use in your answer will include: trustworthy, reliable, team player, organised, creative, responsible and focused. It's not necessary to use all of them, but make sure you get a few of them in your reply. These words indicate the skills that you have, according to what is needed by the employer.

This is your opportunity to make a great first impression and is very important. Try to keep the answer to less than five minutes.

EXAMPLE RESPONSE

"My strong points are that I am focused, organised and a team player. For example, whilst working for my current employer, I was successful in being promoted to the role of chef de partie within 12 months of starting in the kitchen. I like to keep a strong balance between my personal and professional life, this helps me to maintain my high standards at work.

Currently I am studying a diploma course to help with my self-development and career progression. My hobbies include playing guitar and cross-country running. I also recently helped with the cooking at a charity function, during one of my days off.

Overall I would say I am a reliable, focused and hardworking person who is always looking for ways to develop."

It is not necessary to go into too much detail, however the interviewer may want you to expand on some points if they need more information.

2. Would you rather work on your own or within a team?

The interviewer is trying to work out if you are a team player or if you prefer working on your own. It may be obvious to say that you prefer working in a team; however there can be positives and negatives to both.

THE WORK IN A TEAM ANSWER
This may seem like the best answer to use, you will be working in a team and you want to come across as a social, orderly employee who will get on well with co-workers. However, this can also look like a sign that you need other people's help to get the job done.

THE WORK ALONE ANSWER
You might think that this option shows you as an unfriendly person, who is unwilling to participate with the team. However, choosing the work alone option will show that you can manage yourself, and don't need a lot of support.

HOW TO ANSWER
There are two ways to go about answering this question. The first is to use positive aspects from both options in your answer. The second is to choose the way you prefer to work, and highlight that you are ok to work the other way if necessary.

EXAMPLE RESPONSE

"I prefer working as part of a team and enjoy how everybody works together to produce a product for the customer. However, I have no issues with working independently, and would have no problem if I was required to do so.

Within my current job, I work in the main restaurant as part of the team. However, on occasions I am asked to work in a smaller kitchen on my own. I always receive good compliments from the customers and staff on my food when this is the case."

You do not have to choose 1 option when answering this question. By giving both options, you will show yourself as an adaptable candidate.

3. Can you give me an example of when you have helped someone who needed your support?

How to put together your answer:

- What was the situation?
- Why did you provide the help? Were you asked or did you volunteer?
- What actions did you take to help the person?
- What was the end result?

You need to explain a situation where you have helped one of your co-workers to get a job done when they have been struggling. Give an example of where there was a positive outcome, due to the actions you took in helping the other person. For example, did you stay for extra hours when your shift had finished, or offer support when you should have been on a break?

EXAMPLE RESPONSE

"In my current position as a chef de partie I work the starter section as part of the kitchen team. The restaurant has recently employed a newly qualified commis chef from the local catering college and he was assigned to my section to learn how to run the starter section. Once he has the training needed and a good understanding he should be able to run the section comfortably. The commis chef (Darryl) had many good skills but he needed to adjust to the fast pace of the professional kitchen environment, this can be daunting to newly qualified chefs.

I worked closely with Darryl giving him tips and advice on how the section should be run and he was learning quickly. I trained him well and after about six weeks I felt he had the experience and training to be able to run the starter section. Darryl was able to work the section well on quieter days.

However I had a bank holiday weekend booked off for some time and it was going to be Darryl's first busy period running the section whilst I was on my four day break, Friday, Saturday, Sunday and Monday. Well, on the Friday evening I had a phone call from the head chef telling me that Darryl had worked the section very badly and he needed extra help, he had been very slow and even walked out of the kitchen at one point due to the pressure. I discussed this with the head chef and I volunteered to come in over the rest of the bank holiday to work with Darryl and give him the extra training he needed. I made a few tweaks to the way he organised himself during the rush periods where he was struggling a lot. The extra training I gave him really helped him and now he is confident to work the section on his own, even covering my two week holiday period and has since become an invaluable member of the team. I took the next busy bank holiday weekend off and Darryl ran the section very well, being complimented by customers and other chefs in the brigade."

Avoid giving a generic answer, as this will come across as weak. Giving a generic answer will generally give a less positive ending to the answer.

4. What do you enjoy doing outside of work?

This question is asked to find out more about you as a person. Generally it's ok to be honest about your hobbies and interests. Try to demonstrate that you are a well-rounded person with other interests. However, be careful not to mention any hobbies or interests that will be alarming to the interviewer. For example, don't tell them that you like going out and getting smashed a couple of times a week, or that you like to gamble all of your salary away at the casino after work. You don't want to come across as someone who is irresponsible and reckless.

5. How do you handle being criticised by your managers?

You need to look at this question in a positive way, although you do not want the interviewer to think you are a pushover. The question needs to be answered in a way that shows you can accept criticism, but also in a way that shows you are sensitive to it. You do not want to come across as angry, defensive or arrogant. The best response will turn the negative situation into a positive one.

EXAMPLE RESPONSE

"I understand that criticism is necessary and that there has to be somebody setting the standards to be met and making sure that they are. Criticism is all part of the job role and I welcome constructive feedback from others, especially managers with more experience than myself and I respond in a positive way. If I do get criticism from a senior member of the team I make a point to ask questions when I need to, so that I understand exactly what is required of me. I welcome criticism because it leads to better performance and then I know I am doing my job correctly, if I am doing something wrong I want to know.

For example; Whilst working in my current position, a situation arose where I had cooked a fillet steak for a customer's order incorrectly and I was criticised by the sous chef for the mistake. Fillet steaks are an expensive ingredient and wastage is very bad for the business's profits. I felt disappointed in my own performance, I knew that it was important to take the feedback on board and improve my cooking skills in this area. I asked the sous chef questions on how I could improve my cooking of a fillet steak and he gave me some constructive feedback and suggested resting the meat for a longer period of time and adding more seasoning. I took his points on board and made sure I rested the next fillet steak I cooked for a longer period of time and added more salt and pepper. Consequently I received a number of compliments from customers on the fillet steaks I had cooked them, with this dish being the most expensive on the menu I was very happy with my performance and so was the head chef."

 This is a solid response as it has taken the mistake, shown how the criticism was taken on board and how the skills in this area have developed.

6. What motivates you to do a good job?

The interviewer is trying to establish the key to you being successful in the position you are applying for; there is no right or wrong answer to this type of question. This will also show the interviewer that you are aware of what you value in a job.

HOW TO ANSWER

Go through your CV and highlight the tasks that you felt the most motivated to do and enjoyed the most. Make sure the highlighted tasks are related to the role you are applying for.

Be prepared and practice all the questions well. It's best to answer this question (as with all the interview questions) by giving an honest assessment of yourself, but also relating this to the position and the company you are applying for.

7. What do you find demotivates you at work?

The key here is to research the company and the position as much as you can, and identify what the employer is looking for. For example, if a job posting is asking for a person who works well in a large team, and you tell the interviewer that you prefer working in small teams, this won't look good. As mentioned before, you need to be prepared for this type of question. To prepare, make a list of the keywords in the person specification and job description that the employer is looking for, and use those keywords to your advantage. For example, if the employer is looking for a creative person, a good response will be that you find an uncreative workplace demotivating. Turn a negative into a positive.

Always respond to the questions honestly, it's ok to tailor your answers to what the interviewer is looking for, but do not lie.

8. What are your strengths?

When answering this question, you need to give specific strengths that you have, and also provide evidence for these strengths, using examples. You need to provide details of where you have used these strengths before. Try to pick the strengths that are listed in the job description, and use these in your answer.

Try not to just say:
"I am a hard working person who works well under pressure."

Even though there is nothing wrong with this answer, it is not the best answer we can give the interviewer. As I mentioned before, we need to give a specific answer that has evidence to back it up.

EXAMPLE RESPONSE

"I have the ability to work to a consistently high level in a challenging work environment. For example, whilst working for my previous employer I was running the sauce section in the kitchen, at a time when we were aiming for an award. The standards were high and this needed to be maintained at all times. When we achieved the award the business became extremely busy and I worked well, consistently keeping my standards high in a pressurised environment."

Give a specific example of previous experience showing your strengths, and provide detailed evidence to back this up. Always use your own experiences and give an honest response.

9. What are your weaknesses?

This question needs to be answered very carefully. It is usually asked towards the end of an interview. You do not want to say that you don't have any weaknesses, but you also don't want to give a weakness that will damage your chances of having a successful interview.

STEPS TO ANSWER THIS QUESTION:

- Choose a real weakness.
- Choose a skill that relates to the job.
- Explain how you have taken action to improve.
- Finish your explanation strongly with positive results.

EXAMPLE RESPONSE

"Sometimes I can be slightly too honest when giving feedback to co-workers. I am a very straightforward person and to the point. A co-worker gave me some negative feedback on this, so I asked other people how they felt. Most people do not have a problem with this, but I have found that some co-workers do. Following this feedback I decided to change my approach and have since had positive feedback on the change."

Do not choose a weakness that will retract from your ability to do the job. Choose a real weakness that is related, but not associated with a major skill required with the job role.

10. What is your current or previous position, and what were your responsibilities within this role?

When asking this type of question, the interviewer wants to find out about your background and what kind of experience you have. This question will help you to expand upon your CV and clarify any outstanding issues from the interviewer's perspective. Keep your response to this question relevant to the role being applied for. Use the keywords and key phrases that you have identified in the job description and person specification, and use these in your answer.

For example, if the job description responsibilities include "assisting in the monitoring of food quality and presentation" or "ensuring temperature readings are taken and recorded accurately", detail how you have managed these responsibilities in your current or previous positions.

A strong response will show how you have performed the responsibilities in the job description, and times when you have gone above and beyond these responsibilities. Keep your response relevant to the keywords and key phrases. Emphasise how you have successfully completed tasks that are relevant to the job description and how good results were achieved. Identify the top responsibilities in the job description and demonstrate the experience you have in these areas.

11. Why did you leave or are looking to leave your current/previous job?

By asking this type of question, the interviewer is looking for any red flags or areas that are concerning. For example, if you got fired or did not get along with your co-workers. It is crucial when answering this question that you do not criticize your current or previous employer in any way. You do not want to display any bitterness towards your previous employer or any managers that work there. This will only come across negatively, and put the interviewer off hiring you.

Ideally, you should try to be positive in your answer. For example, you could tell them that: I feel I have achieved all that I can at the company, and am now looking for new challenges. The interviewer is looking to see you as a positive person, who doesn't hold grudges towards their previous employers.

EXAMPLE RESPONSE

"Although I have enjoyed working in my current position, I feel it is the right time to take on new challenges. I have worked hard and have learned a lot from the job. The company has treated me really well. I am now in a position where I want to take on new challenges. I have good memories of working at the company, but feel that now is the right time to move on. I have researched your company and the job role and feel I have a lot to offer, starting with my experience and enthusiasm. I know I would be a great addition to the team."

A solid response will demonstrate your ambition to develop as a chef, and your belief that the company will help you to do so. Never put your previous employers down, be positive at all times.

12. What do you know about our company, and why are you interested in working for us?

This is a common interview question, and therefore is one that you need to be well prepared for. This question needs to be answered carefully. You don't want to respond with an answer such as, "I have dreamed of this position all my life."

You need to give a response that is honest and individual, so that you stand out from the other candidates. You also need to show that you have carried out plenty of research on the company. The majority of applicants for the position will be applying for many jobs at the same time, so will not have done any real research on specific companies. You should also consider adding in the key skills you have, which relate to the job role.

EXAMPLE RESPONSE

"I understand that the position will require excellent creative and team working skills. I feel that I am very strong in both of these areas, and that this would make me an outstanding addition to your team. Whilst applying for this position, I have conducted intensive research. Through this, I have noticed a common theme in the company, in that your business is extremely professional and forward thinking. I have also talked to a few members of current staff, who have given me excellent feedback. I am very enthusiastic to be a part of this team and the skills and experience I have will allow me to make a positive impact. I am excited to have the chance to contribute in helping to reach the company's goals."

A solid response will show that you have researched, have specific knowledge about the company, and have an interest in them.

13. What do you see as the key responsibilities/duties of this job?

The interviewer will be looking for your answer to be based around the job description. You need to use the keywords that you have identified in the job description.

A strong response will show the interviewer that you have a full understanding of what the job role responsibilities are. The keywords and the key phrases will need to be identified, and used to prepare a response to answer this type of question.

For example, if the responsibilities in the job description include "ensuring all cleaning plans are followed and hygiene is checked daily" or "making sure that all foods from the chefs section meet the menu specifications", then an answer that demonstrates a comprehensive understanding of why and how these are important, will go a long way to securing the position.

EXAMPLE RESPONSE

"I have studied the job description and person specification carefully. I believe I have a full understanding of the job role and responsibilities. For example, I understand that all foods have to meet menu specifications and company guidelines, and deliver this in my current job role. It's essential to make sure all foods are on point, and follow the Health and Hygiene systems set out in the company policies."

For a strong response, research the job description comprehensively. Make sure you give specific examples of the responsibilities and where the responsibilities have been met in your current job role.

14. How do you feel about doing certain tasks/duties?

The interviewer may give an example of a non-position related task or extra hours that need to be covered.

The interviewer is asking this type of question to check your flexibility within the job role. When responding to this type of question, you should try not to appear stubborn or uncompromising.

Chefs need to be flexible. There could be times when extra hours need to be worked, or staff absences such as holidays will need to be covered. A strong response will show that you are willing to go above and beyond your job responsibilities. It will show that you understand how being flexible in the job role is essential for the company to operate successfully, and continue to develop. Avoid using negative body language when responding to this question, and stay relaxed and confident.

EXAMPLE RESPONSE

"In my current position we have had a very busy period over the summer. I have worked extra shifts and taken on extra tasks, such as helping the pastry chef when my service had finished. I also covered the pastry chef during her holiday period. I am happy to take on these tasks when required during the busy periods, and have no problems with this at all. I fully understand that for the company to operate successfully, there might be times when I will be required to cover tasks or extra hours that are not listed in my general responsibilities."

This is a good response, as it shows an understanding of the fact that flexibility in the job role will greatly benefit the company. It also gives a specific answer to this question.

15. Where do you see yourself in 5 years' time? What is your ultimate goal and how will this position help you to achieve this?

The interviewer is asking this type of question to establish whether you see yourself working long term for the company, how ambitious you are, and what training they will need to provide you with.

Being realistic, you might not know where you are going to be in five years' time, as things change and you may want a new challenge. However you need to show the interviewer that you are planning to stick around for a long period, and that the position is a good fit for your long term plans. In your response, you need to come across as positive about your future, but not overly confident or arrogant.

EXAMPLE RESPONSE

....

"My goal right now is to find a company where I can continue to grow, and I believe this position is perfect. If successful, I would firstly learn all that I can about the role, and would always be looking to further develop my skills and knowledge. Essentially I want to be respected by my co-workers for doing a great job, being professional and being a reliable employee. I am also looking to enter a NVQ programme to further my skills and knowledge, and work towards a promotion within the company. It is important to me that I develop personally and professionally during this time and that I can build on my career."

This is a good response, as it tells them that I am looking to stay for a long period, and that I want to learn the job role and become competent at it.

16. Did you take a culinary course at college? What did you like best and least about the experience?

This type of question is usually asked due to a limited amount of work experience.

Avoid being negative about areas of the course, that are now relevant to the position you are applying for. For example, don't say "I didn't like cleaning the scales from fresh fish" when you are going for an interview in a fish restaurant, or "I didn't enjoy making the fresh pasta" when fresh pasta is a key element of the menu of the company you are applying for. You should even avoid, "I didn't enjoy all the health and safety instructions and cleaning we needed to do", as this is a key responsibility in every workplace.

Identify the key areas in the job description and use these in a positive way. It's ok to say you didn't enjoy an area of college, but be sure it is not an essential job responsibility or a listed personal specification of the job you are applying for.

17. How important is it for you to be clean and tidy, and have an organised workspace?

In this type of question, the interviewer is looking for you to demonstrate your organisation skills, so it is very important to show that you keep yourself organised, clean and tidy. You need to show not only how important keeping your workspace clean and tidy is, but also how you organise yourself overall.

When working as a chef, it is essential that you work in an organised, clean and tidy fashion. Your section needs to be as safe as possible, and must adhere to health and safety laws. Be sure to use the key phrase "clean as you go" in your response.

EXAMPLE RESPONSE

"In my current position it is very important for me to keep my work space well organised at all times. I use the clean as you go method. Keeping myself clean and tidy is essential for me to work efficiently, and to the best of my ability. I keep the quality of my work to a high standard and never sacrifice this for quantity. Keeping clean and tidy also helps me to stay organised, and take a structured approach to my work."

This is a good response, because it shows how important keeping clean and tidy is to you. It also shows that the person is well organised, has a structured approach and can work efficiently.

18. What systems do you follow to make sure your tasks are completed on time?

To give a strong response to this type of question, time management skills need to be highlighted. The systems that are crucial in time management will include: delegation, prioritizing tasks, setting goals, staying organised, minimising stress, meeting deadlines early or putting off a task that needs completing.

This question is asking **what systems** you use to manage time, so giving an example of when you have managed your time is not the best response.

A strong response will show that you can cope with different tasks successfully, understanding exactly what needs doing and that you can be trusted to take initiative for your work.

EXAMPLE RESPONSE

"When I am finishing a shift, I will prepare a list for the following day – my to-do list. I will write this list on a piece of paper, which I will then use to prioritise and work out the order to complete tasks in. If I need to add a task to my list when I am working from it, I will re-order as necessary. This results in completion of my tasks on time."

This is a good response because it shows the system that is being taken, shows organisation, the prioritising of jobs and the ability to set goals to get tasks completed on time.

19. In your opinion, what should the relationship be between the kitchen brigade and the restaurant team?

In this question, the interviewer is looking for your communication skills and your knowledge of the job. Communication between the kitchen brigade and the restaurant team is absolutely crucial in professional kitchens.

Communication needs to be effective in order to deliver all of the orders. Usually, the head chef will be communicating with the kitchen team and the restaurant team whilst running the pass. The pass is the main point where the communication happens between the kitchen and the restaurant. It is also where the organisation of the orders takes place, to get the food out to the customers efficiently. This involves effective communication with the restaurant team. It is very important that the restaurant staff communicate the needs of the customers to the chefs.

To give a strong response, emphasise the importance of communication between the teams in promoting effective teamwork. Give a specific detailed action that you have taken, to communicate effectively with the front of house. Be honest in your response, and make sure you use your own experiences in the answer.

20. What actions do you follow to ensure a safe working environment within the kitchen?

As a professional chef, you should be extremely aware of health and safety and the impact that this has on you and your co-workers. Health and safety in the hospitality industry is governed by the Health and Safety at Work Act 1974.

The kitchen can be a very dangerous place to work, and everybody needs to be aware of hazards and potential hazards at all times.

A professional kitchen has many systems to follow, to create a safe working environment. Everybody needs to have an understanding of how health and safety impacts upon themselves and each other, and an understanding of the procedures that must be followed. Obviously, it's also essential that they can actually follow them too. As a chef, it's imperative that you are aware of what constitutes a safe working environment.

Actions to be followed may include: taking care of your own personal safety and the safety of others in the team, reporting anything that could cause an accident, completing temperature control logs and understanding COSHH and HACCP.

EXAMPLE RESPONSE
....

"*Everybody follows the actions laid out in the Health and Safety At Work Act 1974. Safe working practice is all about keeping safe and promoting a safe working environment. This includes making sure that all foods are prepared and cooked following strict guidelines, that foods are stored correctly, that protective clothing is used when using dangerous chemicals and wet floor signs are put out in the event of a spillage. All these actions should be followed on a day-to-day basis, resulting in a safe working environment.*"

This is a strong response, because it emphasises a good understanding of a safe working environment. It also gives specific actions that are taken, in order to follow these systems.

21. What experience do you have in the purchasing or ordering of stock, and how do you practice stock control?

When asking this question, the interviewer is looking for an indication of your previous experience in this area. The overall responsibility for the ordering and stock control falls on the head chef. This person will need to be aware of the prices of items, costs involved and the budgets that need to be adhered to. It is essential not to over or under order stock, and minimising waste is also important in controlling costs.

However, all chefs need to practice stock control in the kitchen. This generally means using the older items first, and keeping all stock in date to minimise waste. Section chefs will usually check their own stock on their specific section, and provide an order list to the senior chef. The senior chef will make sure all orders are taken for the whole kitchen on a daily basis, and check other areas including the dry goods, cleaning materials, and equipment.

Your response needs to give specific details of the experience you have with purchasing and ordering stock, and stock control. Check the job description to find the requirements of the job, and use these keywords as a blueprint for a relevant answer. All chefs have a responsibility for stock control, and a specific example should be demonstrated in your answer.

22. Tell Me About A Typical Working Week For You?

In this question, the interviewer is looking for you to discuss what you do during a working week – in detail. When you are planning for this question, make sure that you detail areas within your current or previous jobs that relate to the position you are applying for. The more you can relate your previous experience to the position you are applying for, the better your answer will be to this question.

Stay focused on work when answering this question. Do not tell them about other activities that you do on the company's time. For example, don't tell them that you are regularly late due to morning workouts at the gym, or that you will need to finish early regularly to take your children to the swimming pool. Take this opportunity to show the interviewer that you are a well-organised person and an efficient worker.

A strong response will give specific details of your working week, and show that you are focused on work. You need to demonstrate that you are an organised and efficient person. Check the job description to find the requirements of the job and use any relevant keywords in your answer. For example, "The first thing I do on a Monday morning is to check all of the function sheets relevant to my section, for any updates".

23. How important do you feel that it is for a kitchen to maintain temperature control procedures?

Chefs have to be aware that temperature control is very important in the production of food. They must have a good knowledge of the guidelines and systems that are in place to serve safe food. Harmful bacteria are a hazard in every kitchen. They grow rapidly at room temperature, can't be seen and are impossible to physically remove from food. Temperature control is used to control the number of bacteria.

There are two crucial ways to achieve this. Bacteria can be destroyed and reduced in number by cooking and reheating. The growth of bacteria can also be controlled via refrigeration, freezing, cooking, hot holding, cooling and reheating.

ESSENTIAL TEMPERATURES TO USE IN A STRONG RESPONSE

Fridges	Below 8°C, although below 5°C is recommended.
Freezers	-18°C or below.
Cooking	Above 75°C.
Hot Holding	Above 63°C.
Cooling	Cooled as quickly as possible and refrigerated.
Reheating	82°C by law but a core temperature of 70°C for 2 minutes is ok.

A strong response will demonstrate an understanding of all the temperature control methods and the temperatures required to achieve this, also giving an example of **"CRITICAL LIMITS"**. This is a crucial area of all kitchens. You should also give a specific example of how you have followed these systems in your current role. If your current kitchen sets all cooking temperatures to 75°C and all fridges set to 5°C or below, these are called the **"CRITICAL LIMITS"**.

A solid response will show that you have a good knowledge of health and safety issues.

24. What action would you take if there happened to be a discrepancy during a stock take, involving a section you have been working on?

The interviewer asks this type of question to check your problem solving capability, and your ability to set controls.

POSSIBLE CAUSES OF A STOCKTAKE DISCREPANCY
- Deliveries not being checked correctly.
- Wastage: when preparing or cooking, incorrect stock rotation.
- Not following recipes precisely.

POSSIBLE SOLUTIONS FOR A STOCKTAKE DISCREPANCY
- Checking all deliveries correctly.
- Being aware and cutting back on wastage during preparation/cooking.
- Follow recipes precisely.

In this type of question, you need to show your problem solving skills and how you would set a control to stop the problem reoccurring. Be specific in the response.

FORMAT OF THE RESPONSE
- What was the situation?
- What action did you take to find the cause of the discrepancy and set a control?
- What were the specific actions you took?
- What were the positive results?

A strong response will show that the issue has been identified, the obstacles to solve the problem have been identified and that actions have been taken to implement an effective solution.

25. What action would you take if you handled a delivery that wasn't up to the company's standard?

Generally, the interviewer asks this type of question to check your problem solving capability and your ability to set controls.

Chefs need to have good problem solving skills and the ability to set and follow controls. This type of question can be difficult to answer if you do not have any experience in this. However, a strong response will give specific details of experience related to this question.

FORMAT OF THE RESPONSE

- What was the situation?
- What were the items of stock that were below standard?
- What were the specific actions you took?
- What was the positive result?

EXAMPLE RESPONSE

"In my last position I had to handle the fish delivery on a Wednesday. I checked the fish delivery, ticking it off on the invoice sheet. I then noticed that a salmon fillet had been damaged. I phoned Capital, the fish company, about the discrepancy and agreed to send the salmon back for a replacement later in the day. I also noticed that the delivery log sheets were kept in the office, so I suggested to the manager that delivery log sheets were positioned next to the delivery area and helped to implement this. All of the other chefs agreed that my suggestion was a great improvement."

A strong response will show that the issue has been identified, the obstacles to solve the problem have been identified and that actions have been taken to implement an effective solution.

26. What do you understand about HACCP?

HACCP means Hazard Analysis Critical Control Points. This is an essential system, used to help look at how food is handled within the working environment. It introduces procedures to make sure the food produced is safe to eat. There are seven main steps to HACCP:

- Determine problems that could happen (The Hazards).
- Determine the points where problems can happen (The Critical Control Points – CCPS).
- Set critical limits at each CCP. This may include cooking temperature/time.
- Have checklists at CCPS to prevent issues occurring (Monitoring).
- Corrective action to take if something goes wrong (Corrective Action).
- Provide evidence that your HACCP plan is efficient (Verification).
- Keep records of all the evidence (Documentation).

All records must be kept up-to-date and organised. HACCP will need reviewing at set intervals, crucially when something in the food process changes. Advice can be taken from your local Environmental Health Officer.

The food operation must comply with all requirements of current food safety legislation.

EXAMPLE RESPONSE

"In my current company we use this daily, and I follow the Hazard Analysis and Critical Control Point system strictly. This system is used to help with the correct handling of food, to ensure the food is safe to eat. For example, I use the blast chiller to chill high-risk foods such as chicken quickly. The temperature and time is then recorded to ensure the product is safe, along with evidence of the procedure being followed."

A strong response will show a detailed understanding of the Hazard Analysis and Critical Control Point system, with a specific example.

27. What do you understand about COSHH?

By asking this type of question, the interviewer is looking for your understanding of hygiene. You should have a thorough understanding of hygiene when working in a professional environment.

COSHH Means the Control of Substances Hazardous to Health Regulations. The purpose of these regulations is to control an employee's exposure to hazardous substances. These substances include certain chemicals and any other substance hazardous to a person's health.

To satisfy regulations, the employer must assess the dangers present in the workplace that are hazardous to health, and decide what safety precautions are needed. For example, having ventilation when using certain chemicals and using protective clothing.

Employees must be told of the hazards and properly trained. All training needs to be documented.

A strong response will detail a clear understanding of the Control Of Substances Hazardous to Health and how this is crucially important in the working environment. A strong candidate will be able to give a specific example of COSHH being followed within their job role.

28. How do you react when your manager leaves you to work on your job without any supervision?

CHEFS MUST BE RESPONSIBLE

You are responsible for your health and safety, and the health and safety of others. You are preparing and cooking food for customers and there are food safety regulations that have to be followed in order to avoid making people ill. During the chef interview, you should generally expect at least one question based around the theme of responsibility. When answering this, it is essential that you can show that you are a responsible person, who can work with little to no supervision.

EXAMPLE RESPONSE, WITH NO PREVIOUS KITCHEN EXPERIENCE

"In my previous job I was working in my local store on a Saturday and Sunday. I would often be left with lots of responsibility, including stocktaking and ordering of the stock when needed. I followed the systems that were in place and I came up with some improvements on how money could be saved. I trained new staff on using the equipment, and on the systems in place. I work well when I am left unsupervised and I understand the importance of being a responsible person, who can use his own initiative when required."

All questions should be answered using your own personal experiences.

29. Do you have any questions that you would like to ask me about the job?

It's ok not to ask any questions at the end of an interview, although it is recommended that you have a couple of questions pre-pre-pared. Assuming that you have to ask questions at the end of an interview, and asking a bad question as a result, could detract from your ability to reach the next stage.

For example, imagine that your first interview has gone really well. You have responded to the interview questions with consistent and relevant answers, and it's looking good. Then, the interviewer asks the question:

Do you have any questions you would like to ask me about the job?

In trying to impress the interviewer, many people have a list of up to 7 questions that they proceed to ask. However, this is not a good idea. It is acceptable to ask a couple of questions, but asking too many unnecessary questions can damage your chances. It might also give the impression that you haven't listened to what they have said.

POINTS TO REMEMBER
The interviewer may have a busy schedule with other candidates to see, or other work responsibilities. You don't want to take up their time unnecessarily. If you do not have any questions, then it's acceptable to respond with:

"No, thank you. Through the research I have conducted, I have learnt a lot about your company and plenty of my questions have been answered. I realise your time is precious and understand that you are busy. Thank you for interviewing me for this position."

Avoid asking any questions that you should have established during your research. For example,

"Do you offer staff discounts on company facilities?"

"How many holiday days will I get over a year?"

Keep all questions relevant to the position or the company, and don't ask questions designed to try and catch the interviewer out, or to try and make yourself look clever.

EXAMPLE END OF INTERVIEW QUESTIONS

"Through my research, I discovered that your company has gained investors in people status. If I am successful in gaining the position and carry out the job well, are there opportunities to progress within the company?"

"If I am successful today, what are the next steps in the interview process?"

It's great to ask questions at the end of an interview, but make sure you don't: ask too many. You should avoid asking irrelevant questions, which are designed to catch the interviewer out.

EXAMPLE SITUATIONAL INTERVIEW QUESTIONS AND ANSWERS

1. Tell me about a time when you found yourself in a difficult situation. How did you handle it?

In this type of question, the interviewer is trying to establish how you work under pressure, and how you deal with stress in the workplace. A good example response to give would detail a situation where you were challenged, but still achieved a positive result.

FORMAT OF YOUR RESPONSE

• What challenging situation were you in?

• What was the problem you needed to solve?

• How did you resolve the problem?

• What was the positive result?

EXAMPLE RESPONSE

"In my current job, I was managing the kitchen whilst the rest of the team were on a break. A customer was very unhappy with the choices that were available on the menu, and asked the duty manager to see the chef. As I was in charge, I went out to speak to the customer. He was furious! I listened and didn't interrupt and waited for him to calm down. I discussed the situation with the duty manager and produced a different menu for the customer, and he chose the dish he wanted with a couple of changes. I cooked the dish and he was extremely happy, he even complemented my work to the hotel manager. He now does regular business with the hotel. I learnt by carefully listening that I could handle a tense situation and turn a demanding guest into a happy loyal customer."

A solid response will demonstrate your problem solving skills, your initiative, ability to stay calm, and how you can handle a tense situation. The response should be well structured and emphasise the specific action you took to come out with a positive result.

2. Tell me about a time when you noticed an important piece of prep running low, that had been overlooked by the person in charge of this task. And you were working on an altogether different task?

FORMAT OF YOUR RESPONSE

- What was the job you were doing, and what task needed to be done?
- Why had the task been overlooked by others?
- What did you do or say when you noticed the mistake?
- What was the result?

A SOLID RESPONSE

Chefs need to be aware about what is going on around them, and need the confidence to take control of a situation. Be specific in your response. This could involve reporting the mistake to a senior staff member or taking the task on if free to do so, with a positive outcome.

A poor response would generally be where a chef will not take action on the situation, and will overlook the problem. The response will not be structured, and their answer may be generic in quality.

3. Tell me about a time when you took on a task that was not in your job responsibility, without being asked?

This type of interview question is very important. You can demonstrate good motivation, the ability to take action, thinking on your own, and coming up with ideas that help the kitchen develop.

FORMAT OF YOUR RESPONSE:

- What was the responsibility you took on?
- What steps were needed to complete the task?
- What was the specific action you took on, were there any obstacles?
- What was the end result?

A solid response will need to give specific details of where you have taken on a task that is not in your responsibilities, and details of how challenges have been met and overcome. You can show off your motivation and problem solving skills in this type of question. Try to emphasise that you are someone who often takes on non-related job tasks, but only when you are not leaving your own jobs to fall behind.

A poor response will give an answer that does not relate to the question, and will not have evidence of completing a non-specific job responsibility.

4. Tell me about a time that you needed to make a product, where you did not have all of the ingredients for the recipe?

Your response to this question can describe how you utilise your organisational skills and always plan your day, so that you have all of the ingredients required for a day's work in advance. However, on occasions, ingredients are not available or a miscommunication has happened during the ordering or delivery process.

FORMAT OF YOUR RESPONSE:

- What was the situation you were in?
- Why did you not have all the ingredients?
- What steps did you take and were there any obstacles in the way?
- What specific actions did you take?
- What was the result?

A strong response will show evidence of your experience and that you can use your initiative. When you are applying for a position, you need to show the relevant experience you have, so a strong response will show that you can adapt the recipe to safely create an alternative if needed to.

A poor response will not give a specific detailed example, and will not provide a previous work-experience example. The response will not be structured and the answer may be generic in quality.

5. Tell me about a time when you have handled complaints?

There will be times when you have received a complaint from a customer or manager, this happens to everybody at some point. Complaints should never be taken personally and you should be aware of how to deal with them. The key areas to focus on to resolve the complaint should include: listening to the complaint, understanding why they are complaining, providing a solution to the complaint as quickly as possible, or offering apologies to the customer. They generally want someone to explain what has gone wrong.

FORMAT OF YOUR RESPONSE:

- What was the situation?
- Why exactly did the complaint occur?
- How did you resolve the situation?
- What was the result?

EXAMPLE RESPONSE
....

"In my current role I was working on a hot buffet lunch in the restaurant, when a customer came to me complaining about his steak. I listened to his complaint and kept eye contact with him. I apologised to him, whilst keeping myself calm, and offered to cook a fresh steak for him to his preference. He was very happy with the new steak, left a really good tip and complemented my customer service skills to the manager."

A strong response to give will follow this format: listen, apologise, take information, provide a solution, agree on the solution and take the action required to resolve the issue, before finally checking that the customer is ok.

A poor response will not give a specific detailed response and will not provide a previous work experience example. The response will not be structured, and the answer may be generic in quality.

6. Tell me about a time when you have helped a co-worker, who needed assistance to complete a task?

Teamwork is essential in kitchen teams, with chefs working closely together in an organised structure to set standards. All chefs must be aware of time limits to get their tasks done. Occasionally, however, team members may need help.

On the occasions where a team member needs help to achieve their tasks on time, it is important to only do this when you have sufficient time to help. You must make sure that your own tasks will not suffer as a result. Be careful to only offer help on tasks that are within your capabilities, and do not go against any instructions given by a senior chef.

FORMAT OF YOUR RESPONSE:

- What was the situation?
- Why did you help the co-worker?
- What was the specific action you took to help?
- What was the result?

A strong response will be structured and give a specific example of how a co-worker has been helped. An understanding of teamwork and organisation will be shown, as well as demonstrating that the candidate's own work was not affected as a result.

A poor response will not give a specific detailed response, and will not provide a previous work experience example. The response will not be structured, and the answer may be generic in quality.

7. Tell me about a time when you have worked in a fast paced environment and how you handled the conditions?

Chefs must be able to work in a fast paced, kitchen environment. A strong response will show that you can work in such conditions. You should also show a strong enthusiasm for teamwork..

FORMAT OF YOUR RESPONSE:

- Where have you worked in a fast paced environment?
- What steps did you do to prepare for the environment?
- What specific action did you take to handle the work?
- What was the result?
- How would you improve your performance?

EXAMPLE RESPONSE

"In my current job, the kitchen runs at an extremely fast pace. The evening service is very busy, and I work well in these conditions. I actually enjoy the fast working environment. I always organise my workspace, and have all my preparation ready on time. I work in an organised way, and keep my section clean and tidy at all times, as well as having a full understanding of timings and what is required. I handle the conditions extremely well, as I am prepared and understand exactly what is required of me, however I always look for ways to develop my performance."

This is a solid response, as it gives a specific example of working in a fast paced environment. Even though some candidates will not have any experience in a fast paced kitchen, a strong chef for the position will always be able to give a specific answer.

A poor response will not give a specific example and will not show how to handle conditions in a fast paced working environment.

8. Tell me about a time when you have had to work with a co-worker and you did not get along, how did you handle the situation?

You may get on with most people the majority of the time, but at some point in your career you will have had a dispute with a co-worker. Therefore it is not sensible to say that you have never had one. A good example to use is a dispute you had several years ago, and specify how it was resolved.

As with all negative situations, you need to show yourself positively, and this is a good way to show off your listening and communication skills. Describe how you initiated a solution by talking to the co-worker to clear the air, this will show your communication and listening skills.

Perhaps you took your co-worker's feedback on board and came to an agreement in order to move things forward. This is a good way to show that you can resolve issues with co-workers without involving senior managers.

FORMAT OF YOUR RESPONSE:
- What was the situation?
- Why was there an issue?
- What actions did you take to overcome the issues?
- What was the result?

EXAMPLE RESPONSE
....

"Last year I was working in a kitchen, when a new co-worker started. Unfortunately, we did not click at all. One day we had a dispute and the situation became uncomfortable for both of us, and other members of the team. I arranged for us to have a meeting the following day, away from distractions, to clear the air with a view to possibly becoming friends; or at least working together amicably. We resolved our differences and our misunderstanding, and we now have a good working relationship. It's only natural that we can't get on with everybody, but we should always put our differences aside at work and maintain our professionalism."

9. Tell me about a time when you have changed your methods, due to feedback from someone else?

Chefs will be receiving feedback from supervisors in the kitchen on a day-to-day basis, with the aim of maintaining kitchen standards and continuously improving the food that is going out to the customers. Within many professional kitchens, the chefs will be provided with training, both in-house and via courses outside of work. An important part of the learning process is being able to take feedback and improve as a result of it.

To give a strong response, you need to give a specific example of where feedback has been taken on board and where an improvement has been made.

FORMAT OF YOUR RESPONSE:

- What skill needed developing?
- What was the feedback you received?
- What steps did you take to develop your skill?
- How did you specifically do this?
- What was the result?

When you are preparing your response for this type of question, take into consideration that it needs to be specific, giving an exact example. Your response also needs to show that you understand the importance of feedback and how it helps in your professional development.

10. Tell me about a time when you have dealt with a junior member of the team, who was not performing up to standard?

The interviewer asks this type of question when looking for an example of your personnel management skills, and how you manage staff.

This is a good question to show that you are a responsible person, and have management capability. You don't need to be in a senior position to manage staff. For example, you may be working with an apprentice chef who only works on the weekends. If you notice that this person is underperforming, then you need to talk to them calmly and privately, and turn the situation into a positive.

Consider that maybe they are just doing things differently, and not wrong. Make sure you are not angry or irritated, and proceed to talk to them calmly and with care. Let the chef choose their own way of doing the tasks, and ask a couple of questions to see how they feel they are working. When they answer with a good answer, let them know that it's good. If not, give them some practical advice and show them how you would take on the task.

FORMAT OF YOUR RESPONSE:

- What was the situation?
- How was the behaviour inconsistent with the team's standards?
- Why was the co-worker working in this way?
- What action did you take when you noticed this performance?
- What problems did you have?
- What was the result?

All chefs will need the confidence to approach junior members of staff when they see them not performing to the kitchen/company's standards. You must be able to guide the team member towards the correct course of action.

11. Tell me about a time when you were under pressure, and how you coped with the situation?

It is essential to be able to handle pressure and stress when working as a chef. Pressure comes when we are being pushed to complete tasks that we find demanding, in set time frames. Stress is a negative reaction, and occurs when we are put under more much pressure than we can cope with.

A strong response will show that you are able to recognise pressure and stress, understand the impact that both of these have, and that you can deal with pressure and stress appropriately. Working a service in a busy restaurant comes with huge amounts of pressure. Therefore, it's imperative that you are equipped to handle this.

A strong response will also give a clear picture of how well you work in stressful situations. In your response, avoid giving examples of times when you have put yourself in a pointless situation. For example, if you missed some tasks out in your planning and now had to make up the time quickly. Focus on a time when you were given a tough task to complete and how you achieved this successfully.

Another tip is not to focus on how stressed you felt, but to acknowledge that while there was a stressful situation, you coped with this extremely well.

FORMAT OF YOUR RESPONSE:
- What was the work you were doing?
- What caused the pressure you were under?
- How did you deal with the extra pressure?
- What was the result?

EXAMPLE RESPONSE
....

"My current job puts me in a naturally pressured environment. Whilst I generally thrive under this, there are occasions when there is a

higher amount of pressure than usual. A good example of this is on occasions when we have had large bookings in the restaurant that were not made clear to the kitchen beforehand. This put a lot of extra pressure on myself and the team. I react to the situation rather than reacting to the stress, this enables me to focus on the task I need to do. I believe that my ability to organise myself and communicate effectively with other members of the team helps to reduce my stress. The result of this is that I am able to complete pressurised tasks successfully. I am also aware that there is only so much you can reasonably do. In this situation, the key is to stay focused and not to panic, to be able to produce your best."

This is a solid response as it shows where a pressurised environment has been worked in successfully. It also shows an understanding of how to deal with stress.

A poor response will not give specific details, and will not provide a previous work experience example. The response will not be structured and the answer may be generic in quality.

General Chef Interview Questions

- Do you have any holidays pre booked?
- What salary are you currently earning per year?
- What salary are you looking for?
- What is the notice period of your current job?
- How will you be travelling to work?
- What goals have you set out to achieve in your next position?
- Why are you looking to leave your current place of work?
- What type of food are you cooking in your current/previous kitchen?
- Can you describe one of the dishes that you have currently on the menu?
- What is the current kitchen brigade like?
- What is the current structure within the kitchen brigade?
- What role do you have during an evening service?
- How would your current head chef describe you?
- How would your current team members say about you?
- Give an example of when you over-delivered for the company?
- Tell me about the good points within your current place of work?
- Do you have any long-term goals and what do you want to learn in the next 5 years?
- Do you have a current food hygiene certificate?
- What current level of food hygiene certificate do you have?
- What training do you have and do you have any qualifications in cooking?
- Are you looking to take a qualification here or further the qualification you already have? (NVQ)
- Do you have a first aid certificate and is it in date?

- Do you keep yourself physically fit, as there can be a lot of hours standing and physical work?
- If offered the position, when can you start?
- Tell me, why did you decide to become a professional chef?
- Have you applied for other jobs recently and what were they?
- How does this chef position sound to you?
- How would you describe your current head chef?
- What is your best working environment?
- Give me an example of a typical week in your current or previous job role?
- In your current job role, what do you enjoy best and enjoy the least?
- Tell me a single thing you would most like to change about your current role?
- If you were having difficulties, how would you go about asking for help?
- Tell me about the positive personal qualities you will bring to the team?
- Why do you want to be a commis chef?
- Why do you want to be a chef de partie?
- Why did you choose to become a chef?
- Are you a team player?
- What job standards are important to you?
- How did you prepare for this interview?
- How do you establish what's most important when organising your time?
- What action have you taken over the last 12 months, to improve your skills and knowledge to help develop you as a chef?
- Do you have a favourite chef? If so, who, and why do you admire them?
- When do you feel happiest in your job role at work?

- Tell me about the time you worked the hardest and had the greatest feeling of achievement?

- If you were overseeing a function for 150 people, what type of dishes would you put on the menu?

- Can you tell me why we should hire you as the next member of the kitchen team?

Some Reasoning Behind A Selection Of The Questions

WHICH AREAS DO YOU THINK CAN BE IMPROVED WITHIN THE CURRENT KITCHEN SETUP?

This type of question will be put to applicants who are applying for an internal promotion. You will be assessed on your ability to make internal observations within the working environment.

HOW DID YOU HEAR ABOUT THIS JOB ROLE?

The interviewer has asked this question because they want to track the company's advertising campaign. They are also checking to see if the job has been referred to you from a current employee, which may result in a bonus as part of an employment scheme.

DO YOU HAVE ALL OF THE NECESSARY VISAS TO WORK IN THIS COUNTRY?

The interviewer has asked this question to see if they will need to process a work permit.

ARE YOU ABLE TO BE FLEXIBLE WITH THE HOURS AND SHIFTS YOU ARE WORKING?

The interviewer wants to make sure that you are flexible with your working hours and with the conditions of your employment. The interviewer is looking for you to show that you are able to work extra hours during busy periods and that you will be open to changing shifts at short notice. Occasionally you may need to change shifts or work extra hours which does happen due to holiday cover, sickness or during busy periods. When answering this type of question it is important to remain relaxed and do not become tense in your verbal answer or tense in your body language.

WHAT SALARY ARE YOU LOOKING FOR?

The interviewer wants to make sure that you are in the salary expectation range of the company. The interviewer will also see how flexible you are in this range.

WHAT WOULD YOUR REFEREE SAY ABOUT THE WAY YOU WORKED, BASED ON YOUR PREVIOUS POSITION?

The interviewer will ask this type of question to find out how you see yourself.

WHO CAN WE CONTACT FOR REFERENCES?

The interviewer is looking for character references, make sure they are good references and that you have asked them to be a referee before you use them.

Food Related Questions In The Interview

The interviewer may also ask some food related questions. These questions should be based around the style of food that the company is cooking, and there could be 3-5 questions. The questions may include:

- Job specific technical questions.
- Ingredients questions.
- Dishes questions.

An example question might be:

Tell me how you would go about planning a menu. What are the key areas that you look for in a good menu design?

What Happens Next?

To gain the job that you have applied for, there are some essential steps that you need to take. The key here is to practice your skills repeatedly, until they become strong. Practice the questions over and over, and develop your interview skills and technique by using role-plays. You can ask a friend or a family member to sit as the interviewer and take a mock interview. How well you do in an interview will all depend on how much preparation and practice you have put in. The questions listed on the previous pages are all great examples of the types of questions that you'll hear in an interview. Remember also that all of your responses should be based on your own experiences.

More Interview Tips

There are several key steps that you can take to increase your chances of success in your next interview. Be well prepared, practice your interview technique and practice answering sample questions. Some interviews can be casual, but look at this as an opportunity. You need to take advantage of your preparation time. The more preparation you do, the more confident and fluid your responses will be. It is perfectly normal to be nervous before an interview, but remember that the employer wants to see you because you have impressed them, so believe in yourself.

COME ACROSS POSITIVE
You need to have the confidence to expand on any information you have given as evidence in your CV.

RESPONDING TO DIFFICULT INTERVIEW QUESTIONS
When responding to difficult interview questions, use the S.T.A.R method to give a structured fluent answer. Practice your interview technique and practice answering the sample interview questions.

CLOSING ON A CONFIDENT NOTE
Leave a positive impression when the interview is coming to a close. Smile and thank the interviewer for the opportunity. Tell them that you look forward to hearing from them and leave confidently.

RESEARCH THE COMPANY

Do as much research as you can, go through the website thoroughly and check for search results on the Internet. Check the guidebooks for information on the restaurant standards, and use sites such as trip advisor to gain information.

FIRST IMPRESSIONS

First impressions are very important, look smart and professional. Take a copy of your CV with you and plan your route well, to prevent any delays.

GOOD BODY LANGUAGE

Your body language can be rehearsed beforehand. A good idea is to practice in front of the mirror, and then use this good posture in the interview.

Fit In With The Company

Your task is to get the employer to envision you in the job role by the end of the interview, and be comfortable with the possibility of you working in the kitchen. Concentrate on yourself and don't worry about any other candidates. You need to gain the respect of the interviewer, and show that you are a serious candidate from the very start. Be aware that the company probably has a salary range for the position. If this is not in your range, do not proceed with the interview. This information should be found out in your job/company research phase.

When To Apply For Promotion

When you feel that you have enough relevant experience, it's a good time to apply for promotion. However, be aware that a commis chef with experience in a high-end restaurant should only apply for a chef de partie job in another high-end restaurant. If the commis has no experience in a high-end (Michelin star) restaurant, it's generally a waste of time. Your level of skill will need to match the quality of the establishment.

CHAPTER 9

The Chef's Working Interview (Stage)

What is a stage?

A stage is a term for a chef's working interview, this usually entails working in the kitchen for 2-3 hours and preparing a dish for the chef. This is usually the second step in the chef interview process, and if the chef likes you, he will ask you to attend a stage. A stage can take place in several ways and will give you the opportunity to impress the chef. It will also allow you to get a feel for the kitchen.

The four main types of stage are:

- The Mystery Basket.
- Classic Trades Test.
- Working on the section with the chef who runs the section.
- Working on an event (Maybe a wedding function).

CLASSIC TRADES TEST

A classic trades test will require you to recreate a classic dish. This is usually a French dish.

WORKING ON A FUNCTION

Working on a function or during a service for 2-3 hours will give you a feel for the kitchen, and show whether you will fit in.

A MYSTERY BASKET

This chef skill test involves the chef giving you a selection of mystery ingredients. You'll then be asked to cook. During this test you will usually have the kitchen to yourself, and will be able to use other ingredients in the kitchen. There are several ways of doing this, but be prepared to cook a three course meal within a 2 hour time limit.

Tips for a mystery basket

- Research and be aware of all the menus.
- Get an idea of the dishes and ingredients used in the menu.
- Get an idea of the ingredients already in the kitchen that you can use.
- Always cook in the style of the restaurant.

WORKING ON A SECTION

This type of stage will require the chef who is currently running a specific section to work with you on a service for 2-3 hours. You will be able to get a feel for the kitchen, and the chef can see how you work in the team.

Tips for a section stage

Get a copy of the menu straight away and identify all of the foods that will be prepared on the section. Bring your own knives. If you need to ask questions then do so, but don't hold up the process. Work in an organised and efficient manner, taking notes as you go. Above all else, the most important thing to remember is that you should never walk out on a working trial. This is the worst possible thing you could do. Being stressed is ok. This is perfectly natural, and if you need to take a breather then take one. Walking out, however, will show that you are a weak person who is unable to withstand the trials and tribulations of the kitchen environment.

Key Tips For A Successful Working Trial

Have a good knowledge of the restaurant and menus. Also understand the style of restaurant, so that your dishes will match the current dishes. Treat the stage like an interview, dress smartly and have clean, good quality chef whites and shoes.

MUST HAVES IN YOUR EQUIPMENT ROLL

The tools to have with you include: a serrated knife, chopping knife, paring knife, boning knife, scissors and a sharpener. You should also have a sharpie marker, recipe book, and a thermometer; as well as having all of your equipment labelled.

POINTS TO REMEMBER

Always keep yourself calm and make sure you execute any tests or working interviews in the most professional way possible. Show your teamwork skills too. If you have cleaned down your sections, then ask other members of the team if they need any help, be nice and smile. The mentality of a professional kitchen is based on teamwork, so try and act like a part of the team straight away. Be confident in making a connection with the team and offer help wherever possible. Take notes and make sure you understand everything that is being asked of you. Take any feedback from the team about your working trial on board, and above all be positive. If it doesn't work out the first time, persevere. Eventually it will, as long as you keep on improving.

WHEN THE WORKING TRIAL HAS FINISHED

Once the working trial has finished, you will sit down with the chef and they will give you some feedback on your dishes and/or your working performance. Take any criticism positively and get as much feedback as you can from the chef. Ask the chef how you can improve in specific areas.

CHAPTER 10

How to Research The Job Description

How To Research The Job Description And The Person Specification

To prepare for the interview, you will need to research the job description and the person specification. Most employers will use these two important documents to tailor your interview questions. The key here is to be prepared for all the areas covered in these documents. I have provided a sample professional chef entry-level job description and the key evidence areas to focus your preparation on.

Providing evidence on where your experience fits the job description is very important.

A JOB DESCRIPTION FOR A COMMIS CHEF – MAIN DUTIES

Our company requires an energetic and professional employee who has an excellent employment record within the hospitality industry. The candidate will need to thrive in a fast-paced working environment, must have knowledge about food and must have great customer service skills. It is also essential to have excellent communication skills, in order for you to be able to build a relationship with co-workers.

Main job responsibility and requirements:

- Good skills with verbal and written English.
- Minimum of 1 year's experience in a 4/5 star establishment, or a confident, high achiever who can learn quickly.
- A genuine interest in food and drink.

- Problem solving skills.
- High standards of communication and presentation skills.
- Preparing ingredients for senior chefs.
- The prep, cooking and plating of basic dishes.
- To help all other chefs when required.

Key evidence areas to be prepared for:

- Provide evidence of being a professional employee with experience working in a fast paced environment.
- Communicate effectively throughout the interview and be prepared.
- Give evidence of working in a 4/5 star establishment, or that you are a fast learner with high standards.
- Give examples of where you have solved problems related to the job.
- Be ready to communicate effectively and be dressed to impress.
- Give examples of where you have worked with senior chefs effectively.
- Give examples of dishes that you have prepared, cooked and presented, a portfolio of your previous work will also work well.

Applying Through An Agency

There are recruitment agencies around the country, that are specifically tailored for chefs. Many professional agencies will require you to go for a formal interview before being accepted by the agency, so be prepared for an agency interview as you would for a job interview. The agency will then find suitable roles for you, and then you will go through the interview process at the specific job's location. When working with an agency, you will be assigned a recruiter to handle your jobs and you will need to communicate effectively with this person. Communication with your recruiter is needed to gain feedback from your interview. This will include your feedback on how things went, and any client feedback.

CHAPTER 11

Top 10 Insider Tips and Advice

Having worked for a considerable amount of time within professional kitchens, working with many different types of chef, I have noticed various themes that are common for chefs to succeed. I have named these common themes as my top 10 insider tips and advice.

TIP ONE

BE MENTALLY AND PHYSICALLY FIT

It is a big advantage to be physically and mentally fit within the professional kitchen. There can be a lot of lifting to do (this needs to be done correctly to prevent back injuries). It can be stressful at times during busy periods, so good mental health is a must. Cutting out alcohol, caffeine and nicotine from your diet will really improve your mental fitness; and eating at least 5 portions of fruit and vegetables a day will have a great positive effect on your energy levels. Remember that chefs will be standing the whole day, so a good pair of shoes is also needed.

Having too much to do in a small timeframe is a common cause of stress. If tasks are not a priority, don't take them on. Learn to say no to unimportant tasks and delegate tasks. Learning to say no will help to reduce stress, and will help to promote self-development and confidence.

AREAS TO TAKE ACTION

Drink plenty of water to keep yourself hydrated (at least 1.5 litres per day) and eat a healthy diet. Cut out alcohol and caffeine when you need to work at your best, during busy periods or when a promotion opportunity might be coming up. Make sure you are getting enough sleep. While this can be difficult, as stress interrupts sleep, performing regular exercise is a great cure for insomnia.

TIP TWO

PREPARATION

One of the keys to your success will be **PREPARATION**. You must be prepared for every stage of your chef training, from when you first send in your application, to when you are running your own section for the first time. Above all else, preparation is the key.

Examples of where you need to be prepared include:

- CV writing, cover letters and job interviews.
- When running a section within the kitchen, you will need to keep on top of work to be done and orders. To do this, you should construct a comprehensive preparation list (also known as mise en place).

AREAS TO TAKE ACTION

The key areas that need to be focused on include: **Being prepared** to submit an outstanding application form, cover letter and CV. **Practicing** the interview techniques and answering the interview questions contained in this guide, and **practicing the skills** mentioned in earlier chapters of this guide.

TIP THREE

LEARN THE CHEF ROLE REQUIREMENTS INSIDE OUT

Learn everything there is to know about the chef position that you are applying for. Gain a copy of the job description and study the content, making sure you understand exactly what it involves. You should also obtain a copy of the personal qualities required for the job role, and crossmatch them with your skills. Be well prepared for a mystery basket working interview, by reading up and learning the menus.

AREAS TO TAKE ACTION

Get a copy of the job description and the person specification if there is one, and research these important documents. Identify the key requirements of the job description and make sure you can answer interview questions that will be based around these areas. Study the menus and the style of food served at the establishment.

TIP FOUR

UNDERSTAND AND PRACTICE EQUALITY AND FAIRNESS

Equality and fairness play a big part in today's professional kitchens. Professional kitchens have progressed a long way and will keep on doing so, by educating people in equality and fairness. How people are treated at work, regardless of their age, sex, religious beliefs, sexual orientation or background etc. is immensely important. Within professional kitchens you may have many different nationalities and personalities working together. These people will have to work well as a team, and therefore treating different beliefs and personalities with respect is essential.

Professional kitchens need chefs who not only understand equality and fairness, but also believe in and practice it.

AREAS TO TAKE ACTION

Have an understanding of, and believe in equality and fairness. You need to have a good knowledge of what bullying is, how to identify it and how to take action against it. Nobody should have to go to work and be bullied.

For more information on workplace bullying and harassment go to www.gov.uk

TIP FIVE

ASK QUESTIONS

"Knowledge is power", the more you learn the more successful you will be.

This guide is not specific to a precise area of hospitality and catering. There are many different types of professional kitchens, all using different techniques and styles of food. This means that there are a mass of questions which it is possible for you to ask. Make sure the questions you are asking have been thought out and aren't too simple. For example, try not to jump in asking a question that is too basic, such as "where do you keep your salad?" (In the fridge). This looks like you can't find your way around a kitchen. Instead a good question to ask could be, "which supplier do you use for your salad, and does your salad choice vary with the seasons?"

AREAS TO TAKE ACTION
Have questions in mind to ask, and practice asking them and thinking

through your question before asking it. It's important to ask questions when you are starting your career in a new establishment, but it is also important to ask questions throughout your career. This will help you to develop personally and professionally. Asking questions will demonstrate that you are an engaged employee and an asset to the company.

TIP SIX

BE PATIENT, GET FEEDBACK AND LEARN FROM YOUR MISTAKES
Everybody makes mistakes, even big ones. These don't have to leave a bad mark on your career. Most mistakes will help with your personal development, and therefore it is important not to overreact to these mistakes. You must be able to learn from them, and accept them. Future decisions are based on moving forward and

not from previous mistakes. If your mistake has lead to somebody getting hurt, make sure you give him or her an apology and don't be defensive. The key to bouncing back is to be action-orientated, and focus on the future.

Mistakes can be a result of a bad system, and the mistake made will highlight a problem with this system. Explaining yourself in a non-defensive way will help to prevent the same mistake happening in the future.

When a mistake has been made, it's essential that you respond quickly, before assumptions start to be made about your competency. It's recommended to get advice from co-workers or previous employers over a mistake, and ask their advice on how you can recover. Once the mistake has been put behind you, concentrate on the future and work harder to achieve a recovery. Demonstrating that you have learnt from mistakes or past negative feedback will be reassuring for your managers.

AREAS TO TAKE ACTION
Accept responsibility for any mistake you have made or have been a part of, and show that you have learned from the mistake. Behaving differently as you move forward, shows that you can handle equally important tasks in the future. Don't be defensive over your mistakes or blame other people, and don't hold yourself back from moving forward because of a mistake.

TIP SEVEN

HAVE A GOOD UNDERSTANDING OF HEALTH AND SAFETY
Health and safety is essential to the chef's role. Professional kitchens can be a hazardous place to work, with heavy lifting, use of hazardous substances, hot liquids, fires, burns, cuts, etc. You need to have a full understanding of the health and safety precautions required in a professional kitchen, and the correct methods of reporting and dealing with incidents.

How to find information about health and safety in the workplace:

- Induction training, staff training.
- Posters and signs around the workplace.
- Line managers, Head chefs.
- Human resource departments.
- Government websites.

The acts include:

- Health and Safety at work act 1974
- Health and safety information for employees regulations (1989)
- Control of substances hazardous to health (COSHH) Regulations (1999)
- Health and safety (first aid) Regulations (1981)
- Manual Handling Operations Regulations (1992)
- Management of Health and Safety at Work Regulations 1999

AREAS TO TAKE ACTION

Make sure you have a basic understanding of how to deal with incidents and are aware of whom to report incidents to. Maintain high levels of personal hygiene, and work clean and tidy at all times.

TIP EIGHT

RESEARCH AND LEARN ABOUT THE ESTABLISHMENT YOU ARE INTERESTED IN

This is important for various reasons. To start with, you may be asked a question on the application form or during an interview about why you want to join this particular company. As you can imagine, when applying for jobs a candidate may apply for many positions at the same time. The interviewer wants to know exactly why you are attracted to their company.

Most establishments have a website. Visit their website and find out about what they are doing, who they are, their menus, any awards they have and what their company ethos is. Learn about the direction in which the establishment is headed, how many properties are in their group and who their directors are. By showing the interviewer this knowledge, you will be displaying a great interest in the company, and that you have made an effort!

AREAS TO TAKE ACTION

Research and have an in-depth understanding of the company you are interested in working for. This can be achieved by going through the company's website and finding all the relevant information possible. Visit the establishment before the interview, go for a meal in the restaurant and check out the food style and service. Ask for some unpaid work experience in the kitchen.

TIP NINE

VISUALISE & SET YOUR GOALS. HAVE TIME FRAMES TO ACHIEVE THEM BY

This is crucial to your success as a chef. Firstly, you need to have a vision in mind of where you want to be, what you want to achieve and by when.

When creating this vision, imagine the furthest you possibly want to go and the highest level that you want to achieve (don't hold back, this is a vision). Once you have created it, construct the goals you need to achieve to make this vision come true, and the time frames needed for doing so. If you miss a goal from your plan don't worry, just keep progressing towards your vision. Write down your plan, read it daily and keep progressing.

TAKE ACTION POINTS

Have a vision of where you want to be and believe in it. Write down your goals to reach your vision and your time frames for doing so on a piece of paper. Read it daily. This plan can be set as your home screen image on your mobile, so that it's easy to read, or written down and kept on your person.

Believe in your goals and be positive. Understand your limiting beliefs, and that these they can be broken. A lot of the decisions that we make that are holding us back, have been programmed into us from a young age. Having an understanding of this and how to break them will give you a great advantage in achieving your goals.

TEMPLATE FOR A VISION STATEMENT

"In five years I see myself...." Where do you see yourself in five years? Where are you working, what is your position, have you achieved your dream job in your dream place, in your dream country? You need something that makes you want to drive your career and jump out of bed in the morning, if it doesn't excite you, it is not big enough. Write down this vision and set goals to achieve your vision. Think big & keep the statement as your daily ritual to read!

TIP TEN

KEEP ON LEARNING AND DEVELOPING YOUR SKILLS

As a chef it is important to carry on learning new things within your job role. Development within your job role will lead to progression as a chef and promotions, and will also provide you with interesting challenges. As you improve your skills, you'll become a bigger asset to the team.

You can keep developing your skills by asking more experienced co-workers questions, and also finding things out on your own. It's good to work alongside more experienced colleagues to see how they function, and pick up some of their skills.

You can receive essential feedback from your senior chefs by asking for advice on how you can progress, and also participating in formal appraisals that review your performance goals and progression

AREAS TO TAKE ACTION

Take courses in areas you are unfamiliar with, such as pastry or bread making, and take courses in cuisines you are unfamiliar with.

This may include worldwide cuisines, which will influence you to create new dishes. Read new books, or buy old books and put your stamp on older recipes. Use social media to follow your favourite chef role models and gain new ideas and tips. Apply to establishments for unpaid work experience during your days off, or use your holidays to gain this experience. Ask your employer if they will provide training courses, where work-based qualifications can be achieved.

For more information visit – www.becomingachef.co.uk

A Few Final Thoughts

Modern kitchens are far more hospitable to chefs who are working their way up the kitchen ladder, than they were in the past. Companies understand the importance of training and giving advice in a professional structured way. This has changed for the better over the last 10 years, and continues to develop.

When working as a chef it is recommended that you work clean, read a lot of books and gain as much experience of different cuisines as possible. Think about what you are doing within your job role and do not rush your work. Speed will come with experience and practice.

When starting in a new kitchen, be prepared to learn a lot, work hard and be dedicated to succeeding in the job role. You will need to have a good attitude to thrive, and if a situation becomes too comfortable, then put yourself out there and try something new. Your confidence will grow as a result.

MOST IMPORTANTLY, ENJOY THE EXPERIENCE!

A chef career is a great choice. If you have a desire to travel nationally and internationally, get paid and experience great food; becoming a chef could be for you.

A Few More Final Thoughts

You have now reached the end of this interview skills and job progression workbook, and hopefully you are now well prepared to start your career in the hospitality industry. Before you start putting all of the things that you have learnt into practice, I have a few final pieces of advice.

When you have taken the self-assessments contained in this guide, you will understand the areas in which you can improve. The assessments are all based around important areas that are needed to progress your career, and develop as a person. There are many opportunities out there to advance your career, in excellent establishments. The people who gain the best opportunities will be the ones who are well prepared and have a belief in themselves that they will succeed, no matter what anybody tells them. The key components to furthering your career are:

1. SELF-BELIEF
Regardless of what anybody tells you, you can move your career forward. Be confident and believe in yourself, if you have weak areas then improve on these.

2. BEING FULLY PREPARED
Prepare for everything, this is the key to your success. Good preparation will improve your performance, generating great results. As the saying goes – fail to prepare, prepare to fail. Well, prepare to succeed! Using this workbook you can gain all of the personal skills you need, in order to make a success of yourself in the industry.

3. SELF-MOTIVATION AND PERSEVERANCE
How much do you want to succeed as a chef? When I started my career, I had a burning desire to succeed and this kept me motivated to learn as much as I could. Nowadays, I keep myself healthy

and perform regular exercise. This is extremely important, as it keeps my motivation levels up. When you make mistakes, learn from them to perform better the next time. You should always be looking to improve on everything that you do. Perseverance is the key to achieving success. With hard work and discipline, you can become anything that you want to be. Thank you for reading; I wish you all the best of luck.